IMAGES
of America

The Verde River
Bartlett and Horseshoe Dams

ON THE COVER: The Bartlett Dam, on the Verde River, could be called "The Dam of a Thousand Days," because the people responsible for it were given just that long to construct it. On May 9, 1939, they beat that deadline. Representatives from the U.S. Bureau of Reclamation, Salt River Valley Water User's Association, and Hilp and Macco pose on the spillway of their creation, dwarfed by the massive 50-by-50-foot Stoney gates that control the river. (Courtesy River of Time Museum.)

IMAGES
of America

THE VERDE RIVER
BARTLETT AND
HORSESHOE DAMS

Gerard Giordano

ARCADIA
PUBLISHING

Published by Arcadia Publishing
Charleston SC, Chicago IL, Portsmouth NH, San Francisco CA

Library of Congress Control Number: 2009943326

For all general information contact Arcadia Publishing at:
Telephone 843-853-2070
Fax 843-853-0044
E-mail sales@arcadiapublishing.com
For customer service and orders:
Toll-Free 1-888-313-2665

Visit us on the Internet at www.arcadiapublishing.com

To Valerie and Karen—you were always my favorite.

CONTENTS

ACKNOWLEDGMENTS

Nothing sickens me more than the closed door of a library.

—Barbara Tuchman

In a time when many archives are finding reasons to erect barriers to access, I was most fortunate to have the help and cooperation of a great many people who made this retelling of history a reality.

I am forever indebted to the River of Time Museum, located in Fountain Hills, Arizona; Debbie Skehen, president; and Judy Confer, executive director. Their mission statement, which could have served as the preface to this book, is the following: "To lead visitors through creative informative exhibits that recount the captivating past of the Lower Verde River Valley, to depict the importance of water in our Sonoran Desert, and to illustrate how water determines the course of history." I must reserve special thanks for Judy Confer, who all but adopted me, as she threw open the doors of the museum's archive to me.

The U.S. Bureau of Reclamation worked as hard to steer me to information on their reclamation projects as they did to construct them. Jon Czaplici of the Phoenix office always found the time and a way to help. I would also like to thank Karen Cowan of the Boulder City office.

Thanks to the following at McDowell Mountain Regional Park: Rand Hubbell, park supervisor, and Jenny Work, interpretive ranger. She took an interest in the project and went above and beyond the call of duty, ushering me through the park even when she should have stayed home nursing her aches.

Patricia Davidson-Peters and Connie Lightfoot shared their time, recollections, and personal photographs. Connie made the history, and granddaughter Patricia preserved it. I hope this book is as special to you as my time spent with you was to me.

Thanks to history's home—the Arizona Room Collection and Phoenix Public Library, and especially Kathleen Garcia and Maria Hernandez.

Thanks also goes to the following: Evelyn Johnson, executive director of Cave Creek Museum; Danny Piacquadio, Harold's Corral; David Cerull, City of Phoenix Water Services; Todd Bostwick, Pueblo Grande Museum, Phoenix; Michele Hughes, FMI Corporation; Eric Church, Bartlett Lake Marina; Michael Mohl, NavSource; Marjorie Green and Theresa Pinta of Archaeological Consulting Services; and Steve Ayers, Camp Verde Historical Society.

Unless otherwise noted, the images in this volume appear courtesy the U.S. Bureau of Reclamation (USBR); the Library of Congress (LOC); the River of Time Museum of Fountain Hills, Arizona (RTM); and the Phoenix Public Library (PPL).

INTRODUCTION

Arizona's 170-mile Verde River forms at Sullivan Lake in Yavapai County, near the confluence of the Big Chino and Williamson Valley Washes. Its name, Spanish for "green," derives from green malachite deposits along its shores. It flows to the Salt River in Maricopa County east of Phoenix.

People we call the Hohokam entered the area around AD 1, developing irrigation on the Gila and Salt Rivers. Approximately AD 700, they entered the lower Verde Valley. Archaeological studies conducted during the 1990s uncovered a more extensive and sophisticated culture than was thought to exist there. Settlements occupying areas along today's Bartlett and Horseshoe Lakes had become somewhat autonomous from those downstream on the Salt River. Nevertheless, they shared many things. Among these was water as the keystone of survival in the desert. As Hohokam populations grew and developed, so did irrigation technology and their dependence upon it. At their height, they cultivated nearly all the land their skills and the environment allowed. The Verde sites featured ancient aqueducts, ball courts, and multilevel compounds. Crops included corn, tobacco, agave, cotton, and squash. They gathered mesquite beans and cactus fruit and traded with Mexico and Northern Arizona. But their civilization declined and collapsed around AD 1450, before arrival of the Spanish. Theories as to why have been fueling doctoral dissertations for decades. These theories include drought, floods, soil salinization, disease, and warfare. Whatever the reasons, water played a pivotal role for the Hohokam civilization. What was true for the Hohokam would be true for all who followed.

Around AD 1300, while Hohokam civilization was peaking, another group migrated east from the Colorado River. Eventually, the Yavapai dominated the Verde River. Unlike the Hohokam, they were hunters and gatherers, leading a nomadic existence. Their first European contact was with the Spanish. In 1583, Antonio de Espejo came; he discovered some silver and copper deposits around present-day Jerome but did not stay. In 1598, Marcos Farfan de los Godos explored the Verde Valley, staking mining claims in the Prescott area. Juan de Oñate made a last visit to the Yavapai looking for a route to the sea in 1604. The Yavapai would not see Europeans in their lands again for over 200 years.

Perhaps the most significant Spanish incursion in Arizona was that of Eusebio Francisco Kino in 1695, bringing ranching to the Akimel O'Odham (Pima). Among the livestock introduced were cattle and sheep. He brought the first sheep for breeding to Arizona, where they have been raised ever since. So successful was he at establishing cattle that he has been called the "first cowboy."

The Spanish largely ignored the Verde; however, it became a destination for the "mountain men," a name given to American trappers and explorers who entered the Rocky Mountains during the 1820s. Though Central Arizona was part of Mexico at this time, men like Pauline Weaver, Kit Carson, James Ohio Pattie, and Bill Williams came for beaver pelts. Largely unmolested by the Hohokam, beaver formed a unique symbiosis with area rivers such as the Gila, Salt, and Verde. Beaver dams slowed river flow, creating large pools and lush riparian environments. Their elimination resulted in increased flow and erosion, which, combined with the introduction of livestock and non-native plants, caused a severe decline of wetlands.

The Mexican War of 1848 saw the Verde Valley become part of the New Mexico Territory of the United States. Later, in 1863, it became a part of the new Arizona Territory, and in that same year, gold discoveries near Prescott ushered in a rush on the Yavapai homeland. Miners and settlers were not welcomed. Fort Whipple was established in Chino Valley, south of the Verde headwaters, to protect them. The following two forts came soon after: Camp Verde (1866), located at the confluence of Beaver Creek and the Verde, and Fort McDowell (1865), situated along the Verde 7 miles above the Salt River. By 1870, the Stoneman Road connected Fort Whipple and Fort McDowell. The Yavapai had been forcibly removed from the Verde Valley by 1875.

John Y. T. Smith secured a contract to supply hay to Fort McDowell. He established his camp on the Salt River where the city of Phoenix is today. Two years later, Jack Swilling dug a canal near Smith's camp. Thus began the reincarnation of irrigation, originally undertaken by the Hohokam, from which the city's name derives. By 1885, the 60-mile-long Arizona Canal was completed, allowing substantial irrigation of the Salt River Valley. In 1889, the canal's builder, William J. Murphy, and Winfield Scott planted citrus groves. Their success is credited as a catalyst for Phoenix's growth. Phoenix became the state's permanent capital that same year. The Arizona Canal illustrated the potential of desert soil if water was available. Pioneers of the 19th century, though more technologically advanced than the Hohokam, suffered from the same limitation—gravity. Lands north of the canal either needed water pumped in or tributaries of the Salt River, like the Verde, needed to be developed for irrigation. In 1889, encouraged by the new orchards, Augustus C. Sheldon, Prosper P. Parker, and Samuel Symonds traveled north of Phoenix through what they named "Paradise Valley" to Camp Creek, which was below the Bartlett Dam site. They surveyed and irrigated a couple of acres with Verde River water. This was the germ of a vision that would become the Rio Verde Canal Company.

The Rio Verde Canal Company was incorporated in 1891 with Sheldon as president and Parker as secretary. Rio Verde investors began filing for Paradise Valley land under the 1877 Desert Lands Act, which allowed settlers to buy land at $1.25 an acre, provided it was irrigated within three years. Plans were to bring irrigation to Paradise Valley via 140 miles of canals, supplied by a 205,000-acre-foot reservoir to be constructed at the Horseshoe site on the Verde. Rio Verde filed a "notice of appropriation" under an 1893 territorial law that allowed private interests to claim rights to flows from sources like the Verde, so long as they constructed—within a reasonable time frame—and maintained a system that benefited the public. While the vision may have been grand, the timing was less so. Rio Verde was privately funded and dependent upon investment to build. That year a depression, remembered as the "Panic of 1893," did nothing to attract investors. The first attempt to award a contract for project construction failed in 1894 because they planned to pay the contractor with company stock they couldn't sell. They tried again in 1895, this time successfully. A Minnesota conglomerate, Langdon and Douglas Grant, began construction and completed a 730-foot-long diversion tunnel and 15 miles of canal, working well in to 1896. By 1897, the project sat silent under the desert sun. The Panic of 1893 and challenges of an unforgiving desert caught up with Rio Verde. They reasserted their rights to the Verde by filing another notice of appropriation in 1898 and reorganizing in 1901 as the Verde Water and Power Company with Augustus Sheldon as president.

The year 1902 was a watershed not only for the newly formed company's fortunes but also for the entire West. Pres. Theodore Roosevelt signed the Newlands Reclamation Act, creating the U.S. Reclamation Service. It provided money and land for reclamation projects throughout the arid West. The impact of federal involvement on irrigation projects was immediate. The Verde Water and Power Company suffered withdrawals of lands in their proposed service areas as the Department of Interior attempted to control land speculation. More acres were withdrawn by the Interior Secretary beginning in 1901 for conservation within the newly formed Tonto Forest Reserve. This included the Verde Company's dam sites. In 1903, encouraged by the Reclamation Service, Salt River Valley farmers organized the Salt River Valley Water User's Association, which was the initial step for Reclamation's "show piece" project—Theodore Roosevelt Dam on the Salt River. A juggernaut called the Salt River Project (SRP) was born. The "water war" over the Verde had begun.

In 1903, Verde Water and Power argued in federal court that the government owed the company compensation for improvements made on the Verde up to 1897. The Reclamation Service's first director, Frederick Newell, called the claim fraudulent, nauseating, and sacrilegious. The judge ruled against the Verde Company, stating that infrastructure built under the Rio Verde had fallen into disrepair and the company had failed to make progress. Though the company lingered on, it ceased being a serious player on the Verde.

By 1914, the Salt River Valley Water User's Association had completed Roosevelt Dam and Granite Reef Diversion Dam, located below the confluence of the Salt and Verde Rivers and asserted the Verde, a tributary of the Salt, was part of their system. They filed their own notice of appropriation for the Verde's water presenting plans to develop the river. However, that same year, a new group, the Paradise Verde Water User's Association, emerged with their own plans. In 1917, this organization filed for their water and rights of way. The Salt River Association objected vigorously, sending a delegation to Washington, D.C., to do so. This began a 16-year struggle to win the Verde.

Wilson administration Interior Secretary Franklin Lane directed assistant C. Bradley to hold hearings to mitigate between the two. The Salt River Association rejected any compromise. The Paradise Valley Association responded by reorganizing under the Smith Act for taxing authority to help build their project, becoming Paradise Verde Irrigation District. Salt River retaliated, approving a $5-per-acre assessment for their Horseshoe Dam. In 1919, Bradley appointed Reclamation engineer Homer Hamlin to investigate if the Paradise Association had a realistic chance of completing their plans. He concluded the Paradise plans were inadequate, unsafe, grossly in error, and criminal. Secretary Lane gave the Paradise Association six months to come up with new plans. Conversely, Hamlin concluded Salt River should have all the rights to the Verde. Meanwhile, City of Phoenix voters, not waiting for the outcome of the struggle, voted for water improvements that would culminate in construction of a pipeline from the Verde to supply municipal water in 1922.

Wilson's next Interior secretary, John Barton Payne, arrived in 1920. Unacquainted with the controversy, Payne started by refusing to go along with the Salt River Association's position simply because Reclamation favored it. The Verde District produced their own report, prepared by respected engineers A. L. Harris and Fred Noetzli, refuting Hamlin's report. Impressed, Secretary Payne accepted this report, which vindicated Paradise Verde while not threatening the Salt River Project's water needs. Harding administration Interior Secretary Albert Fall allowed Paradise to pursue their plans provided they could acquire financing by 1923. They failed at this. That same year, they changed their name to the Verde River Irrigation and Power District and were granted six- and nine-month extensions to secure project money. They did not meet these deadlines. Coolidge administration Interior Secretary Hubert Work, seeing promise in their plans, granted the Verde District until 1925 to sell bonds. They could not. A depressed agriculture market and fear over continued challenges to the district by the Salt River Project scared investors off.

Secretary Work suggested the two sides meet to work out a cooperative agreement to develop the Verde. However, after reaching agreement with the Salt River, the Verde District backed out. Repeated attempts at agreement failed. Without title to lands needed to build dams and canals, nor improvements to support their water rights, Secretary Work concluded the Verde District could not build their project. In 1926, the Senate Committee on Irrigation and Power agreed. In 1928, the district was forced to compromise with the Bureau of Reclamation (as it had become known) and the Salt River Project. But the agreement was rejected by association shareholders in 1929. An ongoing drought had left Roosevelt Lake dry by 1928, and frightened shareholders who felt that all, not just part, of the Verde should belong to SRP were the main reason for the rejection.

This rejection motivated incoming Interior Secretary Ray Lyman Wilbur (Hoover administration) to grant the Verde District's reapplication to develop the river, giving them five years. The Salt River contested this decision, and the battle continued. The Great Depression made private investment impossible, so the Verde District lobbied, unsuccessfully, for federal dollars. The Franklin Roosevelt administration in 1933 ordered another study. The resulting report was the "Preston and Stanford Report." It favored the Verde District, recommending an ambitious incarnation of the project including hydroelectric power capacity to help pay the debt, a storage dam at the Horseshoe site, and a diversion dam at the Bartlett site downstream. Under pressure from Arizona's political leaders to bring jobs, Interior Secretary Harold Ickes announced a multi-million dollar Public Works Administration loan to get work started. It seemed the Verde River Irrigation and Power District had won. But the Salt River Valley Water User's Association had one more card to play.

An ongoing drought had plagued Arizona since 1925. The Salt River Project had invested heavily in hydroelectric power during the 1920s to raise revenue. But the dry weather and a depressed economy killed profits, leaving the Salt River Project barely solvent. The question was raised—if the Salt River Project couldn't do it, what chance would the upstart Verde District have? These considerations, along with constant association lobbying, caused Reclamation Commissioner Elwood Mead to dispatch engineer E. B. Debler for one more report. He concluded that the Salt River Project was far better suited to complete the project than the Verde District. He also recommended that the Bartlett site (surveyed by Verde District secretary William Bartlett) be used for the storage dam site. Commissioner Mead withdrew support for the Verde District. Incoming association president Lin B. Orme immediately requested federal funding for his dam. Politicians from Arizona sensed the change. Congresswoman Isabella Greenway solicited President Roosevelt's support for the project, while Senators Carl Hayden and Henry Ashurst took their cues from Mead's change of heart. Gov. Benjamin Moeur also supported the Salt River Project. The effect was devastating to the Verde District. On October 4, 1934, Secretary Ickes pulled the district's loan. Paradise Valley investors hung effigies of Moeur, Greenway, and Ickes. But the fight was lost. Retaining William Bartlett's name, the Bureau of Reclamation built the Bartlett Dam between 1936 and 1939 for the Salt River Project to operate.

With the sinking of the USS *Arizona* at Pearl Harbor in 1941, America's resources would be tapped as never before for World War II. Phelps Dodge Corporation had copper mines at Morenci in Greenlee County and was directed to increase its production by 80 percent. They would expand Morenci operations using processes that consumed vast quantities of water. That water would come from the Black River, a source of the Salt River. To compensate the Salt River Project for the water withdrawn from its watershed, Phelps Dodge proposed construction of a reservoir at the Horseshoe site on the Verde to replace Black River water. After one unsuccessful vote, association shareholders approved the project in April 1944. The last dam actually constructed on the lower Verde was completed in 1946.

In 1922, several western states, including Arizona, signed the Colorado River Compact, dividing that river's water. Dissatisfied with its allotment, Arizona disputed the agreement. The battle climaxed in a series of court cases known as *Arizona v. California* during the 1960s. In 1968, the Central Arizona Project was authorized by Congress to bring Colorado River water to Central Arizona. After pumping water uphill across 300 miles of desert, the water needed to be stored somewhere. The Orme Dam, at the confluence of the Verde and Salt Rivers, was proposed. It would have inundated major portions of the Fort McDowell Yavapai Nation, which had been created by Theodore Roosevelt in 1903. The Yavapai and a coalition of environmental groups defeated it. The Cliff Dam, proposed above the Bartlett Dam for flood control, was also defeated.

Aptly called "The River of Time" by the museum of the same name in Fountain Hills, the Verde is a time line of Arizona history—so much so that this book limits itself to history of the lower portion of the Verde River, which begins 55 miles upstream from the Salt River at the Verde River Sheep Bridge.

One

Dams before Man

An official associated with the City of Phoenix's Tres Rios Wetlands restoration project considers a "new tenant" who arrived shortly after its inception in 1995. In fact, beavers were once plentiful along desert rivers like the Gila, the Salt, and the Verde, building dams and altering the environment long before people made their mark on the land. (Courtesy of the City of Phoenix.)

Though most people living in the Phoenix area don't know it, towering cottonwood trees, cattails, and lush marshes were the norm on the Verde, Salt, and Gila Rivers. Instead, they think of the dry riverbed (above) that runs through the Salt River Valley today. Trappers had eradicated the beaver by the 1840s, changing stream flows and eroding the marshes. Invasive species such as salt cedar, overgrazing by livestock, and finally, reclamation efforts transformed the rivers. The Tres Rios Wetlands Project (below) is an attempt by the City of Phoenix to restore a portion of the Salt River to its former condition. It is very much how theses rivers looked prior to significant alteration by man. (Both, courtesy of author's collection.)

About a billion years ago, the lower Verde was a seabed. Intense volcanic activity, earthquakes, weather, and uplift would form the river's west flank, the McDowell Mountains (above). By 100 million years ago, the waters had receded, and eastern mountains of the Verde Valley, the Mazatzals (below), began forming. At that time, a river many miles wide drained the area flowing north. This was opposite of the Verde today, which flows south. About eight million years ago, Northern Arizona began to rise, forming the Colorado Plateau, which eventually reversed the river's course. In geological terms, humans were relative newcomers to the Verde Valley, having arrived a mere 8,000 years ago. (Both, courtesy of author's collection.)

The Hohokam entered the Verde around AD 700. The Lower Verde Archaeological Project, undertaken by the U.S. Bureau of Reclamation beginning in 1991, examined two major Hohokam periods, the pre-classic period, AD 500–1150, and the classic period, AD 1150–1450. At least 26 separate settlements had been identified in the areas between Bartlett and Horseshoe Lakes. Some sites, like Scorpion Point, were much larger than anyone had anticipated. Above is Scorpion Point Village, near Horseshoe Lake, around 1991. Interestingly, early Hohokam society seemed to avoid capturing aquatic animals. Later in their history, they hunted beavers. However, beaver populations were never actually threatened during Hohokam occupation. Beaver dams are quite sophisticated, employing arched construction. In addition to dam building, beavers constructed canals. It seems that innovations credited to the Hohokam and to the dam builders of modern times had actually been "patented" by these industrious little creatures long before. (Reprinted with permission of Statistical Research, Inc.)

The investigations of the 1990s revealed a complex culture that was semi-autonomous from the Hohokam who populated the Salt and Gila Rivers downstream. Scorpion Point, a pre-classic-period site, derived its name from a style of projectile point found there that was reminiscent of the shape of that arachnid. (Reprinted with permission of Statistical Research, Inc.)

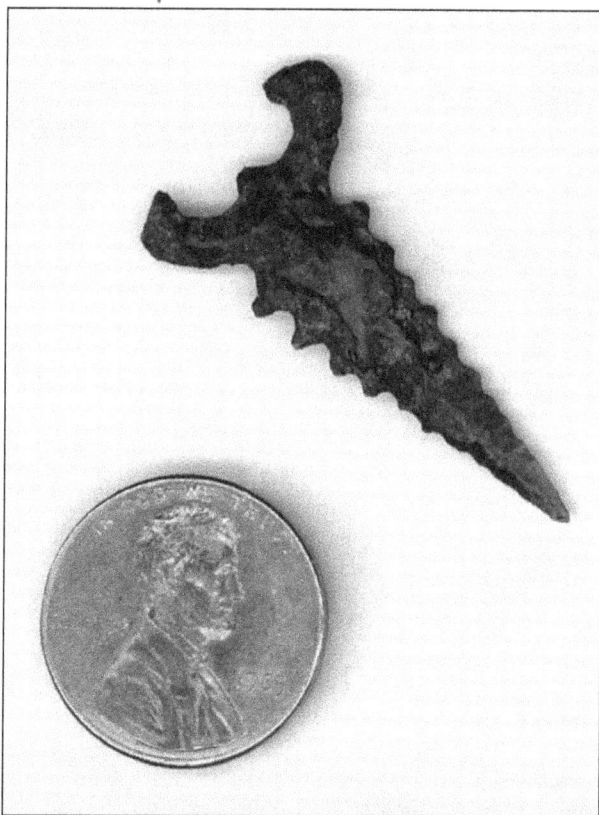

Roadhouse Ruins, pictured here around 1991, is a classic-period site, also near Horseshoe Lake. It was named after a complex that was bisected by a local road. In spite of the invasiveness and partial destruction, sites in the area were otherwise found to be remarkably undisturbed. (Reprinted with permission of Statistical Research, Inc.)

Above is the foundation of a Hohokam pit house at Scorpion Point Village, pictured in 1991. Below is a sketch of what a typical Hohokam pit house probably looked like. The walls were comprised of a mud plaster applied over brush, similar to adobe. Its name derives from the sunken foundation in which wooden poles were set to support the structure. Archaeologists believe that there were as many as 400 such houses within the settlement. A family's pit houses tended to cluster together within settlements, so neighbors were also relatives. (Above, reprinted with permission of Statistical Research, Inc.; below, illustration by Michael Hampshire, courtesy of Pueblo Grande Museum, City of Phoenix, Arizona.)

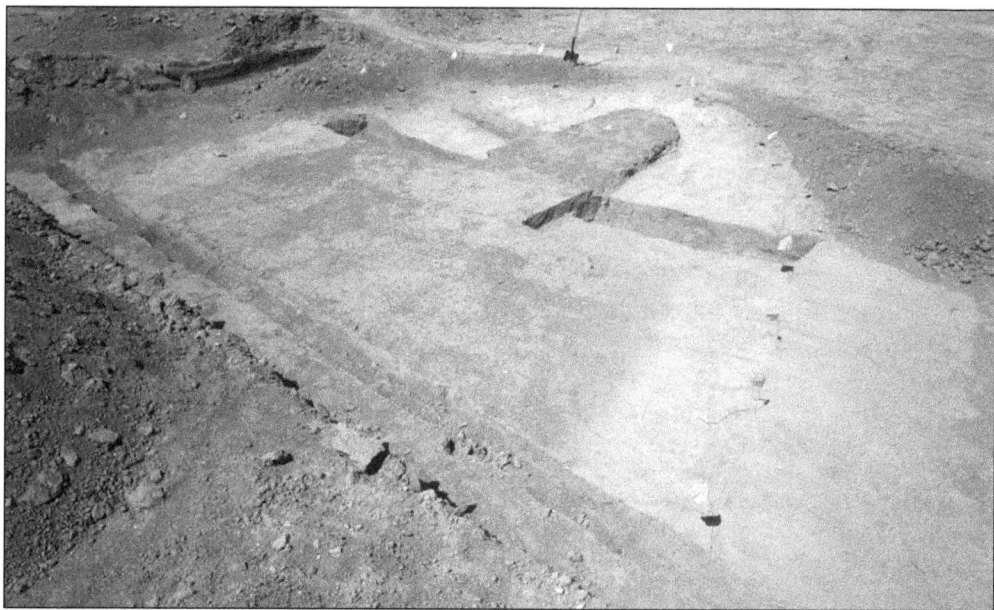

Investigations lasted until 1994. Archaeologists unearthed several ball courts, such as the one pictured above, at Scorpion Point. Shrouded in mystery, they probably derived from similar games played in Mexico, possibly looking like the illustration below. Spanish conquistadores observed a Mexican version of the games, which symbolized the rise and fall of the sun and order in the universe. A ball and goals were employed. Players experienced attention more akin to Roman gladiators than modern athletes. They sometimes stood in for gods; losing team captains could have their heads cut off; figurines of players were found at burial sites. Courts were also used for ceremonies and cremations. Around AD 1200, the Hohokam stopped using them. Many became trash mounds. (Above, reprinted with permission of Statistical Research, Inc.; below, illustration by Michael Hampshire, courtesy of Pueblo Grande Museum, City of Phoenix, Arizona.)

As the Hohokam moved into the classic period, the mud-plaster pit houses were replaced with stacked stone masonry compounds such as the ones found at Roadhouse Ruin (above). This type of construction was organized in the courtyard compounds that are characteristic of the period and looked very much like the artist's conception below. During this time, population numbers were at their height. Land under cultivation was also at its apex, though more crops were being dedicated to food production, such as corn, than commodities for trade, such as cotton. (Above, reprinted with permission of Statistical Research, Inc.; below, illustration by Michael Hampshire, courtesy of Pueblo Grande Museum, City of Phoenix, Arizona.)

As population and cultivation increased, the technology of irrigation grew more sophisticated. One of the best examples of this is located near Mullen Wash at Horseshoe Lake. In order to bring water to the maximum number of fields, the Hohokam constructed aqueducts over the washes. At Mullen Wash (above), the stone wall extending on either side of the wash once supported an elevated canal. Seen at ground level (below) the aqueduct extends across the wash at low water. Considering the propensity of the Verde and its ancillary streams to flash flood, one can only imagine the amount of upkeep and repair that was required to keep the ancient system viable. To date, it is the only prehistoric aqueduct discovered in Arizona. (Both, reprinted with permission of Statistical Research, Inc.)

It is believed that the aqueduct at Mullen Wash was used to support a wooden canal made up of segments of hollowed-out logs called *canoas*, similar to those pictured above. It would not be the last time a wooden conveyance would transport water from the Verde. (Reprinted with permission of Statistical Research, Inc.)

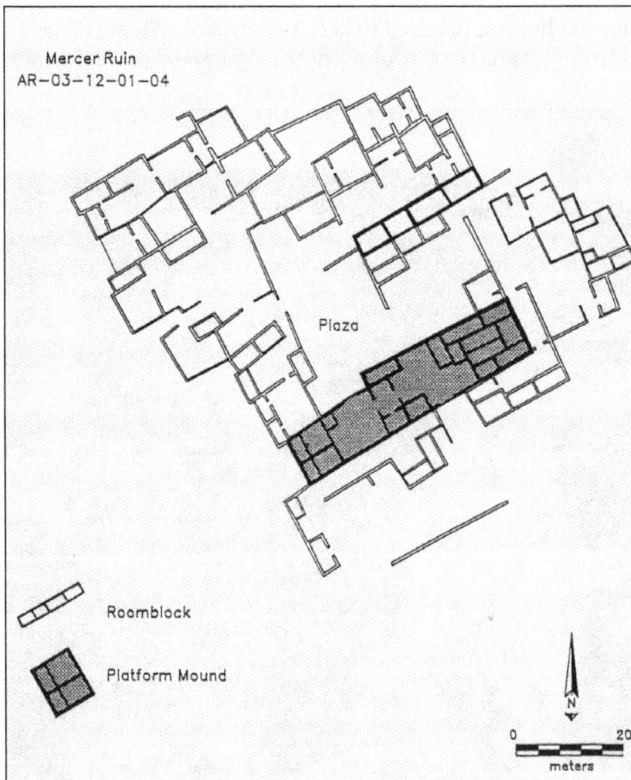

At the height of the classic period, the Hohokam constructed a large compound called the Mercer Ruin, which is now under Horseshoe Lake. Its floor plan is shown at left. Toward the end of the classic period, the Hohokam constructed fortified compounds on hilltops at the periphery of the Salt River Valley. They needed to be on their guard, but against whom is unknown. (Reprinted from *The Bartlett Reservoir Cultural Resources Survey*, Archaeological Consulting Services, Tempe.)

Two

HUNTERS AND
THE HUNTED

Mountain man Christopher "Kit" Carson poses some time after 1860. Seeking beaver pelts, he and a group of trappers ventured up to the San Francisco (Verde) River in 1829. The following is an excerpt from his diary: "We were nightly harassed by the Native Americans who would frequently crawl into our camp, steal a trap or two, kill a mule or horse, and do whatever damage they could." (Courtesy of LOC, cwpbh 00514.)

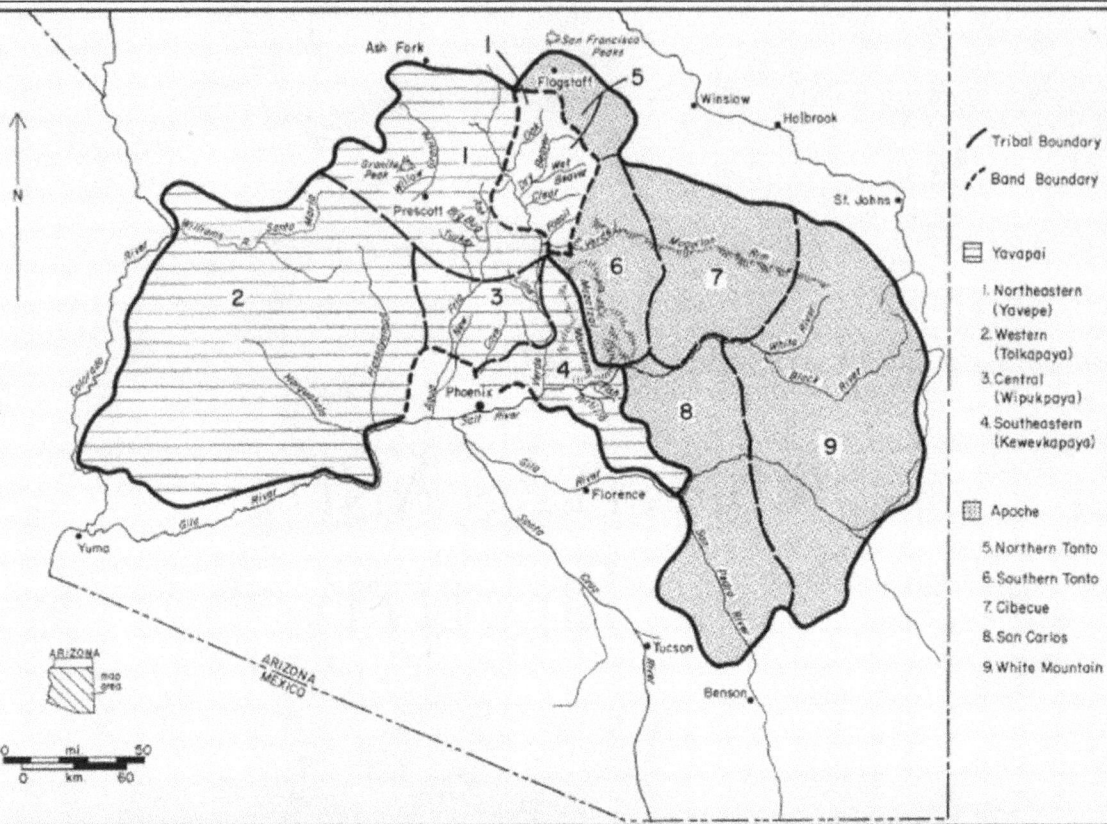

Kit Carson was describing one of the tribal groups that ranged over Central Arizona, the Yavapai or the Western Apache, most likely the Yavapai. The Yavapai were divided into five groups, of which four survive today but with spelling variations. The four groups are the following: the western group, the Tolkapaya; the northeastern groups, the Yavepe and the Wipukpaya; and the southeastern group, the Kewevkapaya. Europeans often confused the similarities in appearance and lifestyles with those of the Apache, who actually arrived much later. Many early narratives refer to Yavapais as "Apache-Mohave" or simply Apache. The following five groups represent the Western Apache: the Northern and Southern Tonto, Cibecue, San Carlos, and White Mountain. Both tribes looked upon the Europeans as antagonists and invaders. The Europeans largely regarded the Native Americans with fear and contempt. Their history became a tragic story of extermination and removal. With their numbers decimated, lands they once ranged over, as seen on the map above, are reduced to the small nations they occupy today. (Reprinted with permission of Statistical Research, Inc.)

Above is an example of a traditional Yavapai shelter called an *oo-wah*, and below are examples of traditional Apache shelters called wickiups. Both have been commonly referred to as wickiups. Both were dome-shaped dwellings made of brush over wood frames. This illustrates part of the confusion surrounding the distinction between Yavapai and Western Apache peoples. The Yavapai language has Yuman roots, leading scholars to think they migrated east from the Colorado River, probably around 1300. The Apache are an Athabaskan-speaking people whose roots are in Canada. They likely arrived in the area during the 1500s. Europeans arriving in the area saw people leading the same nomadic lifestyles. Since anthropology was rarely a concern for them, they simply regarded all of them as Apaches. (Above, courtesy of the National Archives; below, courtesy of LOC, 3c12830u.)

A recurring figure in Arizona's history, Jack Swilling is most known for his canal company that helped found Phoenix. However, his path would cross several times with the Yavapai and Apache with far-reaching consequences. He led a militia in pursuit of Tonto Apaches in 1860. He was instrumental in the capture of the notorious Apache Magnus Coloradas in 1862. His gold discoveries at Rich Hill near Prescott in 1863 helped begin the gold rush into the Yavapai homeland. This caused the army to move in and set up a string of forts along the Verde to protect the miners and settlers. A supporter of the policy of "extermination" of Native Americans he perceived as hostile, his adoption of two orphaned Apaches is an apparent contradiction to his attitude toward other Native Americans. One of the orphans he adopted is seen posing with him in this only known portrait of Swilling, taken around 1875. (Courtesy of McClintock Collection, Arizona Room, PPL.)

Gen. Irvin McDowell, photographed around the time of the Civil War, took command of the Department of the Pacific in 1864. The fort bearing his name was established in 1865, and though he visited in 1866, it's not clearly known if he visited Arizona prior to this. The McDowell Mountains and McDowell Road in Phoenix were named for him as well. (Courtesy of LOC, LC-BH82- 4109.)

Col. George Stoneman, seen here during the Civil War, was the first military commander of the Arizona Territory in 1870. He ordered the widening of a trail that connected two of the Central Arizona forts, Forts Whipple and McDowell. Known as the Stoneman Road, it shortened the distance between the forts, becoming a main route of travel in Central Arizona for 20 years. (Courtesy of LOC LC-B813- 1562 A.)

Civil War–era army ambulance wagons similar to those seen above were a common sight on the Stoneman Road while Fort McDowell operated between 1870 and 1890. They were popular because their rugged construction was well suited for travel over the desert terrain. The military road ensured rapid movement of troops and supplies within the Yavapai lands. The road fell into disuse after the decommissioning of Fort McDowell. Today a 7.5-mile section of the road is preserved within McDowell Mountain Regional Park as the Stoneman Historical Trail and Military Route (below) and is a popular equestrian trail. (Above, courtesy of LOC, cwpb 01986; below, courtesy of author's collection.)

Near the end of its time as one of the most strategically located forts in the Arizona Territory, Fort McDowell is pictured above around 1890. Unlike Hollywood versions of western forts, it lacked a log palisade along its perimeter. Warfare styles of the Yavapai and Apache made an all-out assault on a military installation an unlikely event. Below, two cavalrymen pose in front of a fort building. Apparently, the men stationed there were better soldiers than farmers, as in 1866 an attempt to supply themselves with badly needed fruits, vegetables, and fodder failed. Thereafter, this task was contracted out to civilians. These civilians later founded the city of Phoenix using irrigation and ever-increasing amounts of water. (Above, courtesy of Arizona State Library, Archives and Public Records, History and Archives Division, Phoenix, 96-20270; below, courtesy of RTM.)

Beginning in 1864, an effort was made in the Arizona Territory to concentrate the Yavapai on reservations. Camp Date Creek and Fort McDowell were used first and were followed by other locations within the traditional homeland, which included Rio Verde Reservation near Fort Verde in 1871. The Yavapai and Apache, and even the Yavapai groups, did not always coexist without hostility among themselves. This is one reason why many served as scouts for the army or as reservation police. Above is a contingent of scouts serving at Camp Date Creek; below is a group who served at Rio Verde. (Both, courtesy of the National Archives.)

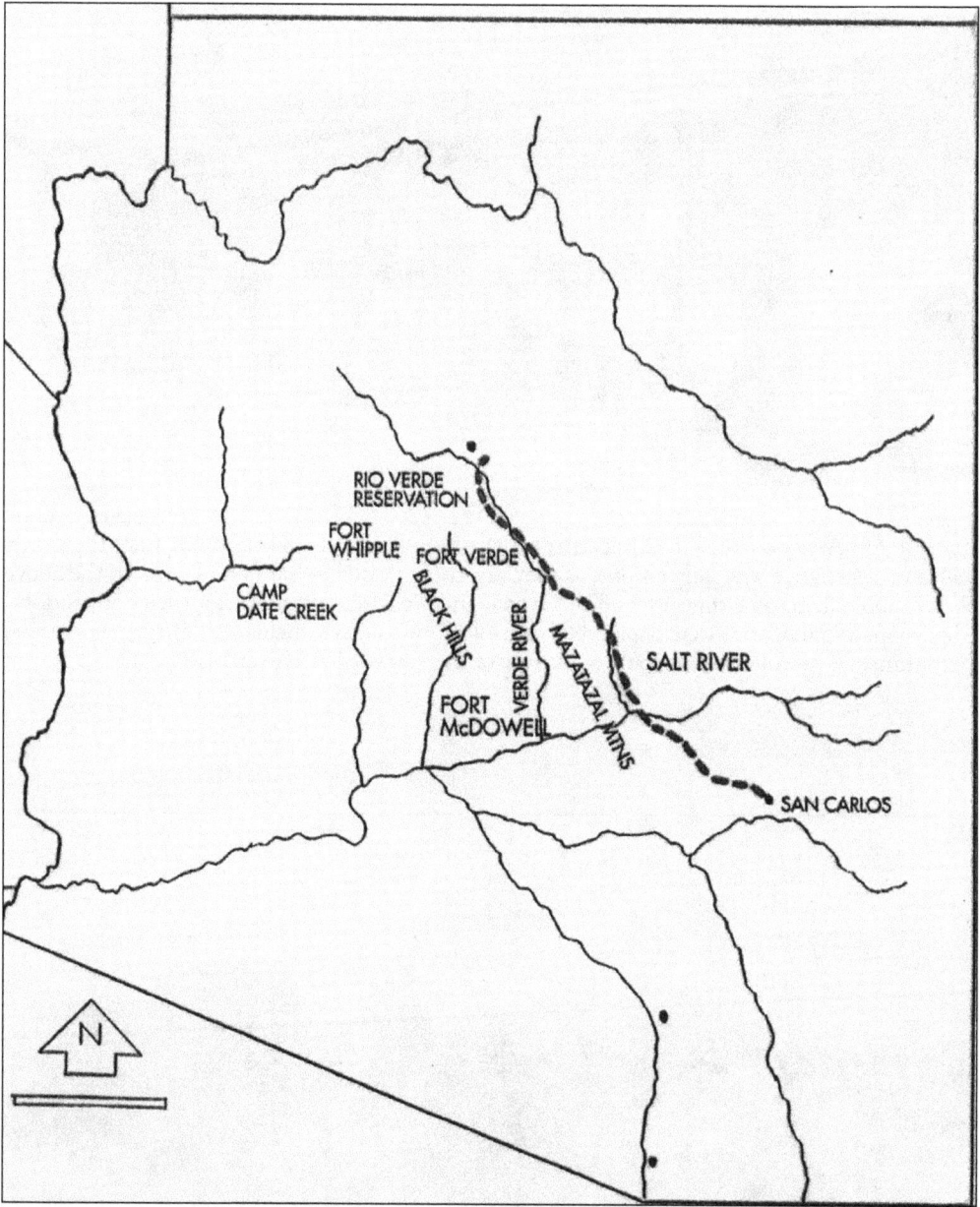

RIO VERDE RESERVATION

FORT WHIPPLE

FORT VERDE

CAMP DATE CREEK

BLACK HILLS

VERDE RIVER

MAZATZAL MTNS

SALT RIVER

FORT McDOWELL

SAN CARLOS

N

While the forts may have been a welcomed sight to the ever-increasing number of settlers moving into Central Arizona, it spelled doom for the Yavapai. Their nomadic lifestyle was vulnerable to the incursion of outsiders. By 1870, starvation, disease, and a policy of extermination had reduced their numbers to less than 1,000. In 1871, Gen. George Crook took command of the Arizona Territory issuing "General Order No. 10." This designated any Native American not confined to a reservation by 1872 to be considered hostile and to be subdued by whatever means necessary. In that same year, a massacre of Yavapai at Skeleton Cave, which is near the Salt River, marked the end of their resistance but not their misery. In 1875, they were to be consolidated along with many of the Apaches on the San Carlos Reservation. A disastrous 180-mile winter march followed. The route, outline on the map above, became known as the Yavapai Trail of Tears. Along its way, more than 100 died. (Courtesy of RTM.)

Yavapai and Apache tribes lived together on the San Carlos Reservation. At first, their only source of sustenance was daily rations, as they were not allowed to hunt and forage. In the above photograph, taken some time after 1875, Yavapai and Apache wait in line for their daily rations. Conditions at San Carlos were deplorable, causing many to leave, including Chiricahua Apache Geronimo and his followers in 1885. (Courtesy of the National Archives.)

General Crook returned to Arizona, enlisting Yavapai help in tracking Geronimo down with the promise that they might be allowed to return to their homeland. This wouldn't occur until many years after. One of those places was Fort McDowell, abandoned in 1890. This photograph dates from about 1930 and shows the crumbling walls of the old fort. (Courtesy of Arizona State Library, Archives and Public Records, History and Archives Division, Phoenix, 97-0651.)

On the right, Pres. Theodore Roosevelt is posing at Roosevelt Lake, Arizona, in 1911. In 1903, Roosevelt became aware of difficulties involving Yavapai leaving San Carlos to settle around Fort McDowell. Whites were threatened by the returning Yavapai. The Yavapai felt they were taking up General Crook's pledge allowing them to return for their participation in the Geronimo campaign. Reservations supervisor Frank Mead was dispatched to assess the situation. He learned that the Yavapai wished only to be allowed to support themselves there. Below is the decree signed by President Roosevelt on September 15, 1903, creating the Fort McDowell Indian Reservation. Other reservations that followed were the Camp Verde Reservation near Fort Verde in 1910, which was shared with Tonto Apache, and the Yavapai Prescott Reservation at the site of Fort Whipple in 1935. (Right, courtesy of LOC, 3a08917u; below, courtesy of RTM.)

EXECUTIVE ORDER.

White House, September 15, 1903.

It is hereby ordered that so much of the land of the Camp McDowell abandoned military reservation as may not have been legally settled upon nor have valid claims attaching thereto under the provisions of the Act of Congress approved August 23, 1894, (U.S.Statutes at Large, Vol.28, P.491), be, and the same is hereby set aside and reserved for the use and occupancy of such Mojave-Apache Indians as are now living thereon or in the vicinity and such other Indians as the Secretary of the Interior may hereafter deem necessary to place thereon.

The lands so withdrawn and reserved will include all tracts to which valid rights have not attached under the provisions of the said Act of Congress and in addition thereto all those tracts upon the reservation containing government improvements which were reserved from settlement by the said Act of Congress, and which consist of (1) the immediate site of the old camp, containing buildings and a good artesian well, (2) the post garden, (3) the U. S. Government Farm, (4) the lands lying north of the old camp and embracing or containing the old government irrigation ditch, and (5) the target practice grounds.

Theodore Roosevelt

Yours in the Cause,
Carlos Montezuma M.d.

Photographed in 1890, Wassaja, translated to "signaling," was born to the Yavapai tribe around 1865. Kidnapped as a boy by the Pima, he would have been sold into slavery—that is if immigrant Carlos Gentile had not bought him to become his son, known as Carlos Montezuma. Montezuma attended the University of Illinois and went on to Chicago Medical College, becoming a physician in 1889. He soon left private practice to work for the Bureau of Indians Affairs at various locations, ending with the Carlisle Indian School in Pennsylvania. In 1900, he travelled to Arizona as a football team physician, and he reacquainted himself with his relatives and roots. Disillusioned, he returned to private practice in Chicago and became an outspoken critic of federal Native American policy. He was ecstatic over the Yavapai return to Fort McDowell but equally dismayed at ongoing attempts to remove them from their valuable land. He became a tireless spokesman not only for the Yavapai but also for other tribes, including the Pima who had taken him as a child. He advocated full citizenship for Native Americans until his death in 1923. (Courtesy of the National Archives.)

Though most of the viable mining claims along the Verde Valley occurred along its upper reaches, in places like Jerome, prospectors, undeterred, flocked to the slopes of the McDowell Mountains in search of gold and other valuable metals. No viable mines were ever established there. Nevertheless, speculation continued into the 20th century. The Dixie Mine was established around 1917. In a rare look inside the mine, seen above in 2009, a deep vertical shaft gapes open from the floor. Below, a mountain of tailings sits in mute testimony to some 500 feet of tunneling that produced very little wealth. Located within McDowell Mountain Regional Park, it is a popular destination for hikers today, prompting the placement of an iron gate at the mine's entrance for public safety. (Both, courtesy of author's collection.).

Pictured above in 2009 is an abandoned 20th-century miner's test hole near the Dixie Mine, which only produces cobwebs today. Below, just feet from it, are centuries-old inscriptions of the Hohokam. Both miners and Hohokam members came to this place for something they valued. We know what the trappers and miners wanted from this place as well as others throughout Central Arizona. The pelts and gold they sought cost the Yavapai and Apache more dearly than any fortunes lost by investors in failed claims. As for the Hohokam, some have speculated that symbols here indicated water, which today floods the shafts of the old Dixie Mine. No one really knows. What is known is that water is the one true constant treasure found along the Verde. (Both, courtesy of author's collection.)

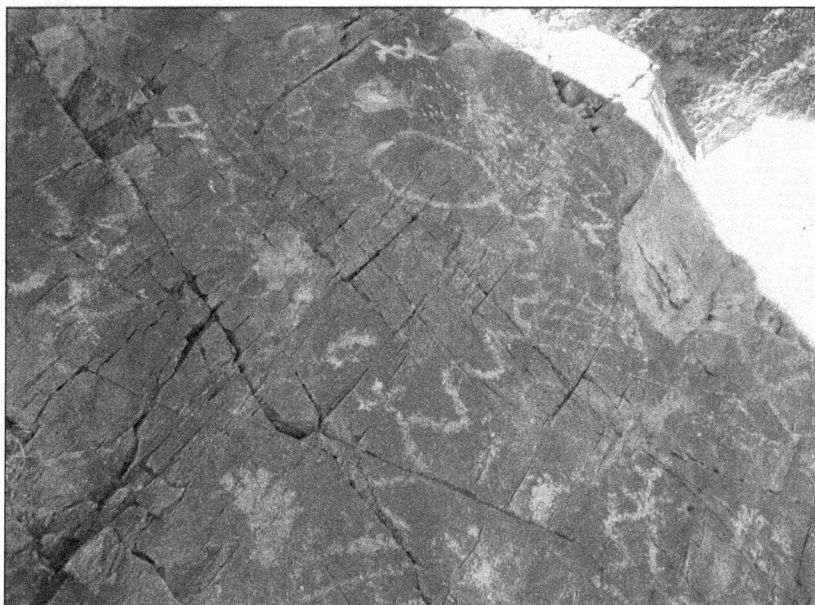

Three

THE HERDSMEN

Fr. Eusebio Kino is immortalized in this statue displayed on the Arizona State Capitol in Phoenix. Known as the "first cowboy," he brought livestock to Arizona, reaching as far north as the Gila River in 1694. However, his impact on ranching was largely blunted by the Pima Revolt in 1751. Meanwhile, fear of Native Americans, the Civil War, and an arid geography stunted ranching efforts in Central Arizona until the late 1800s. (Courtesy of author's collection.)

The end of the Civil War brought people into Arizona. Overgrazing in Texas caused ranchers to move to Arizona, while many displaced by the war also came west to seek a new life, and since the war was over, the army could once more spare the troops to protect them. The land along the lower Verde was a ranching mecca. Around the beginning of the 20th century, cattle ranching reached its peak. The P-Bar Ranch, northeast of Phoenix, was a significant area cattle ranch consisting of 33,000 acres. It was jigsawed together from state and federal leased lands and private property. The map above shows the boundaries of the P-Bar along with information on historic sites and how it evolved. The portion above Thompson Peak Road is the McDowell Mountain Regional Park today. Below that, the ranch developed into the town of Fountain Hills. Within the park boundaries is the original P-Bar Ranch House site. (Courtesy of RTM.)

Pictured are Delsie Dee Bardoll and Richard Robbins at their wedding in 1939. Delsie inherited the P-Bar when her first husband, Lee Bardoll, died in 1938. In 1935, Lee Bardoll acquired the P-Bar from Bill Cole, who inherited it when his father, "Pink" Cole, died in 1926. Prior to that, it was the smaller Pemberton Ranch (the assumed origin of the "P" in P-Bar). Pink Cole bought the ranch during the early 1920s from Henry Pemberton, who made it his homestead some time after 1906. Delsie and "Dick" Robbins ranched at the P-Bar until 1955. Bob Evans, an architect, held the ranch until 1956, selling to Fred Eldean, who increased holdings to 35,300 acres. In 1964, the Page Land and Cattle Company (with Eldean as president), through land swaps, increased the privately owned acreage to 11,380 acres in the southern portion of the holdings. This was deeded to the McCulloch Company in 1968 and was developed into Fountain Hills beginning in 1970. The remaining 21,099 acres in the north became McDowell Mountain Regional Park. (Courtesy of RTM.)

The old P-Bar ranch house (above) is photographed in 1941 with ranch dog Payson in the back a Model T Ford. The three-room house was constructed from local materials. The walls consisted of saguaro cactus ribs, which were plastered with clay. It had an adobe fireplace. The fence was made from live ocotillo canes and mesquite poles. Windows were canvas and frame, though glass was eventually added. The original dirt floors were overlaid with wood. Water was supplied by an artesian well. In 1948, a new ranch house (below, around 1976) was finished. This house, called "Adds Well," featured five rooms and was used by the Robbins family until they sold the ranch in 1955. The old P-Bar house burned during the 1950s. The Adds Well location was demolished in 1979. (Both, courtesy of RTM.)

Dick Robbins shows off his calf roping skills above at the Fort McDowell Rodeo in 1951. He was a charter member of the Rodeo Cowboys Association and often won at these events. (Courtesy of RTM.)

The skills displayed by P-Bar cowboys weren't merely for show. To the right, in 1954, ranch hands tend to cattle at the Windmill Pasture, a 20-acre pasture with a well that was one of the primary watering holes for stock. (Courtesy of RTM.)

Another major cattle ranch on the lower Verde was the 92,000-acre Box Bar Ranch, established in 1915 by W. W. Moore and Frank Asher. It was later sold to Will Ryan in 1954. It was located just north of the Fort McDowell Reservation and extended up to Bartlett Lake. Pictured is the ranch headquarters in 1955. (Courtesy of RTM.)

There was no electricity at the Box Bar; therefore, air conditioning wasn't installed, and the bathroom was located across a pasture and took the form of an outhouse. The lack of luxuries didn't seem to bother Will Ryan's grandchildren, who are seen to the left taking advantage of the ranch house's close proximity to the Verde River in 1954. (Courtesy of RTM.)

While conjuring a romantic image of Arizona ranch life, this photograph taken some time during the 1950s shows more than just Box Bar cowboys herding cattle across the Verde River. Compare this photograph to the restored wetlands on page 12. Along the far shore, cobble from erosion can be clearly seen. This occurred all along the Verde River beginning in the late 1840s after beaver had been eradicated from the river, eliminating their dammed pools, which, in turn, increased the velocity of the river's flow. Unrestricted grazing of cattle beginning in the 1880s further increased erosion, which created unfavorable conditions for the towering cottonwood trees that populated the shore at one time. They are replaced by salt cedar, seen along the shores, an invasive shrub introduced to Arizona in the early 1800s. Today the Box Bar land is the site of three PDCs (planned development communities)—Vista Verde, Tonto Verde, and Rio Verde. Much of the surrounding land has been closed in an attempt to allow it return to a natural state. (Courtesy of RTM.)

Built on the P-Bar Ranch, one of the areas earliest PDCs, Fountain Hills, is seen here in 1971. Beginning in 1970, it was developed by Robert P. McCulloch, the same developer who bought the London Bridge and created Lake Havasu City. Pictured here is the community center piece, a 562-foot-high fountain, which was the world's tallest at that time. (Courtesy of RTM.)

As part of the transactions conducted by Fred Eldean for the P-Bar Ranch, the northern two-thirds of the P-Bar was acquired by Maricopa County, becoming the McDowell Mountain Regional Park in 1964. Within the county park system, it is unsurpassed for its breathtaking mountain vistas. (Courtesy of the McDowell Mountain Regional Park.)

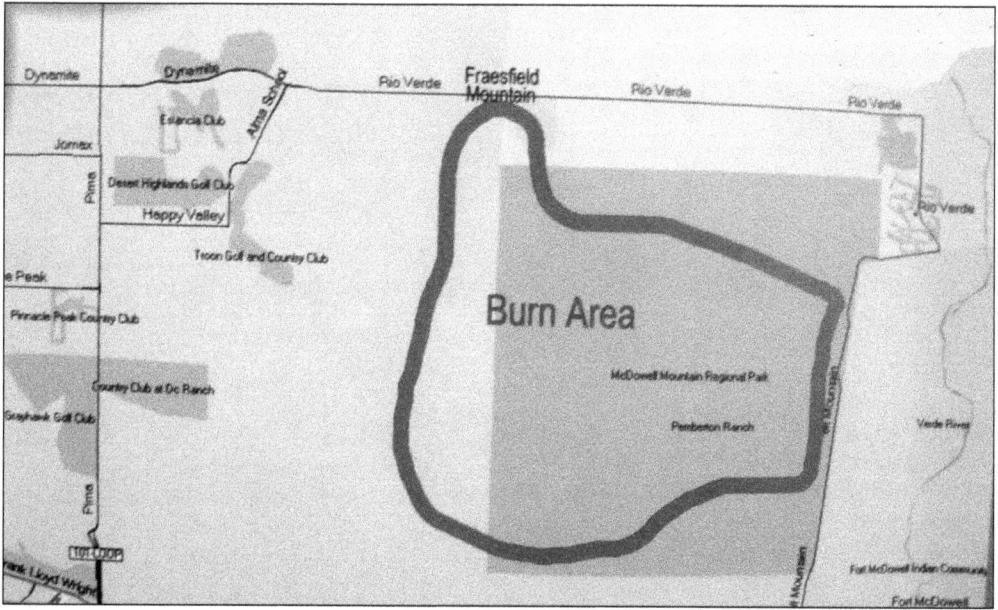

During July 7–10, 1995, lightning was the cause of the Rio Fire, which burned over 22 square miles of Maricopa County, including 14,000 acres of the McDowell Mountain Regional Park. The burn area is shown in the map above. The blaze caused mass destruction of wildlife and habitat, which included many saguaros, some of which were more than 200 years old. Below is the fire as viewed from a park campground on July 8, 1995. Over 300 homes were evacuated, while 459 firefighters fought the conflagration at a cost $1.7 million. Amazingly, though homes in Fountain Hills, North Scottsdale, and Carefree were threatened, no lives or homes were lost. (Both, courtesy of the McDowell Mountain Regional Park.)

For similar reasons as cattle ranching, sheep ranching didn't move into Central Arizona until the late 1800s. Like the one above around 1910, sheep herds came to Central Arizona from Northern Arizona, particularly around Flagstaff. The Atlantic and Pacific Railroad, completed in 1883, served as a catalyst for ranching. It wasn't long before sheep began to spread south, competing for prime range with cattle. Just on the other side of Mazatzals, the Tonto Basin became the backdrop for the bloody Pleasant Valley War between the Tewksburys and the Grahams. This conflict was based, in part, on the tension between cattlemen and sheep men. One of the fears expressed by cattlemen, which came to pass, was the expansion of sheep into the Salt and Verde River Valleys, accelerated by the arrival of the railroad in Phoenix in 1887. One of the greatest sources of conflict rose over the routes taken by shepherds moving flocks between summer and winter pastures. This was resolved in 1916 by establishing a system of sheep driveways over public lands. (Courtesy of the McClintock Collection, Arizona Room, PPL.)

The Forest Service designated driveways and issued grazing permits. Paths between Tonto National Forest allotments and drives between seasonal pastures necessitated crossing the Verde River for 11,000 sheep owned by Dr. Ralph Raymond of Flagstaff during the 1930s and 1940s. Dam construction meant there were fewer places to cross. Above is the Horseshoe Bridge in 1944, prior to its removal for the Horseshoe Dam construction. (Courtesy of LOC, HAER ARIZ, 13-CACR.V, 1-33.)

In 1987, Dr. Raymond's Basque foreman, Frank Auza, stands below the bridge he built in 1944. It replaced the Horseshoe Bridge, which was inadequate for sheep. He chose this site, 3 miles north of Horseshoe Lake, because it was high enough for a bridge to escape seasonal floods. By many, it's considered the dividing line between the upper and lower Verde Valley. (Courtesy of LOC, HAER ARIZ, HAER ARIZ, 13-CACR.V, 1-4.)

The 1943 photograph on the left is of George W. Smith (right) with his son, Les Smith. A Cave Creek carpenter, George Smith was Frank Auza's partner in designing and constructing the bridge. In a procedure that is unheard of today for its informality, Smith and Auza moved forward with the project on little more than a verbal agreement from the Tonto National Forest to build on public land. (Courtesy of LOC, HAER ARIZ, 13-CACR.V, 1-36.)

Auza and Smith visited the Blue Point Bridge, which was built in 1916 over the Salt River, for ideas. Though finding materials was a challenge due to war shortages, they designed a suspension bridge using steel cables salvaged from Mayer Mines, lumber from Flagstaff, and concrete from Phoenix. Below, Les Smith, with no hard hat or safety cable, works on the main suspension cables in 1943. (Courtesy of LOC, HAER ARIZ, 13-CACR.V, 1-37.)

First to cross the newly completed bridge in 1943, Frank Auza puts George Smith's carpentry skills to the test as he trots his horse over the wooden deck suspended over the river below. At low water, the bridge would be about 45 feet above the water line. One flood saw that distance reduced to 10 feet. (Courtesy of LOC, HAER ARIZ, 13-CACR.V, 1-27.)

Sheep mob the 3-foot-wide bridge, which was named the Red Point Bridge, during one of the first crossings in 1944. The name was derived from red-colored rock at one of the abutments. Unfortunately, the name didn't catch on, and the bridge became popularly referred to as the Verde River Sheep Bridge. (Courtesy of LOC, HAER ARIZ, 13-CACR.V, 1-39.)

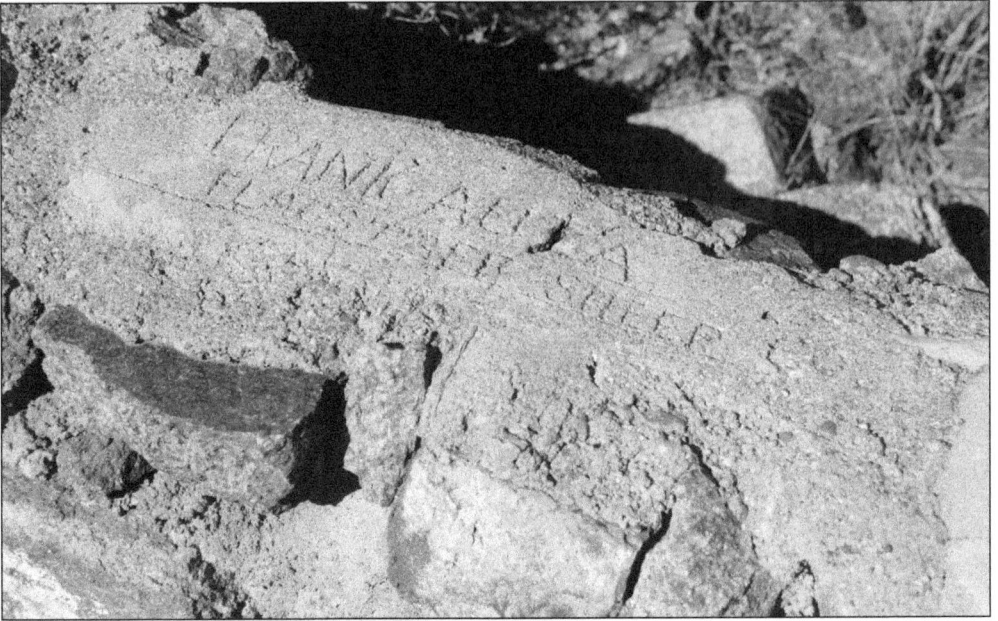

Upon completion of the bridge, it was decided to add concrete buttresses to the towers to strengthen it. This was done in 1944, and all the men who worked on the bridge scratched their names into the concrete. The inscription above reads "Frank Auza, Flagstaff Sheep Co." It was photographed along with the image below in 1987, just prior to demolition of the bridge, which had seen its last sheep drive in 1979. The 691-foot span had fallen into disrepair as sheep ranching declined in Arizona. It was still used by hikers, and the Forest Service, fearing for their safety, took it down. Recognizing the historic significance of the bridge, a copy of it was constructed next to the original towers, which were preserved along with Frank Auza's name. (Both, courtesy of LOC, HAER ARIZ, 13-CACR.V, 1-42 and HAER ARIZ, 13-CACR.V, 1-2.)

Four

A 28-MILE-LONG BARREL

In 1922, two men stand on top of the culmination of a 15-year odyssey by the City of Phoenix to provide its citizens with a reliable supply of water. The result was a gravity-fed pipeline made of redwood staves and held together by steel straps. It was constructed much like a series of barrels connected end to end for 28 miles. (Courtesy of the City of Phoenix, Water Services Department.)

Phoenix was photographed in 1918 looking south down Central Avenue from about Washington Street. The population was nearing 29,000 and growing rapidly, mainly due to agriculture. Water was key to continued growth. In 1907, the city achieved a conversion of the water supply from private to municipal control. One the city's early goals was to secure better-quality water than what was being pumped from its salty groundwater wells. As early as 1906, engineers were looking at the possibility of pumping fresh Verde River water from the Fort McDowell Reservation. In 1915, a study confirmed that a gravity-fed system could be constructed, and the city filed for a right-of-way through the Yavapai lands the next year. After delays caused by World War I, voters approved a bond issue to construct the pipeline. (Courtesy of PPL.)

Phoenix city manager V. A. Thompson is photographed here around 1920. In 1913, as superintendent of the water works, he undertook a personal investigation of the viability of the Verde pipeline proposal. He brought an esteemed hydrologist, Hiram Philips, into the planning process. Philips stayed on as a consulting engineer to help oversee the work. (Courtesy of PPL.)

Phoenix city engineer L. B. Hitchcock, responsible for the cost and execution of the project, was described as having "made a most enviable record in the construction work." The total cost of the project came in at $1,525,282.62. That is roughly $16,381,535.34 in 2009. The authorizing bond election was passed by a margin of 25 to 1, illustrating the value citizens placed on water. (Courtesy of PPL.)

Water intake construction on the Verde River at Fort McDowell was underway in 1920. Water was carried through 12,000 feet of concrete pipe to a sand trap. From there, filtered water passed into the redwood pipe to begin its journey. In December 1921, with the project nearly complete and accident free, the car Hiram Philips was riding in overturned near this location, killing him. (Courtesy of PPL.)

Known as the "Big Redwood Line," the pipe sections came in 36- and 38-inch diameters, which were supplied by the Pacific Tank Company and the Redwood Manufacturing Company. Use of steel or concrete would have been preferred; however, wooden pipelines were not unusual during this time, and redwood was chosen as a cost consideration. Shown here around 1921, the pipeline snakes around a mountain. (Courtesy of PPL.)

Each of the arched wooden staves that made up the pipeline was 1 9/16 inches thick, 5 1/2 inches wide, and 12 to 18 feet long. Half-inch-wide steel bands were bolted around the staves, keeping them together, and were spaced 2 1/4 to 10 inches apart, depending on the pressure exerted on the pipe section. Pictured around 1921, the pipeline extends across the desert toward Phoenix. (Courtesy of PPL.)

Pictured around 1922, the pipeline crosses a canal. In return for allowing access to the Verde and the right-of-way across Native American land, the Yavapai and Pima were trained and employed for the building project. Additionally, the two tribes had free access to water for domestic use through taps in the line. (Courtesy of PPL.)

After crossing the Arizona Canal 5 miles east of Scottsdale, the pipeline followed Thomas Road to a reservoir 6 miles east of Central Avenue. From there, it went on to Sixteenth Street, then south to McDowell Road, and finally west to Twelfth Street, which is where it connected to city mains. The pipeline was buried in some places, as seen here, prior to being completely covered around 1922. (Courtesy of PPL.)

The system was plagued with problems, not the least of which involved cowboys on the range shooting holes in it so they could drink or bathe. The city had to hire someone to plug the bullet holes. The Big Redwood Line only lasted until about 1931. Here a surviving section is displayed at the River of Time Museum in Fountain Hills. (Courtesy of RTM.)

Five

WATER WAR

The Bartlett Dam took 3 years to build and 20 to fight over, as the Salt River Valley Water User's Association fought the Paradise Valley Water User's Association over Verde River rights between 1914 and 1934. The "referees" were six secretaries of Interior. They are, from top to bottom and left to right, Franklin Lane, John Barton Payne, Albert Fall, Hubert Work, Ray Lyman Wilbur, and Harold L. Ickes. (Courtesy of LOC.)

Pride of ARIZONA

GROWN BY INGLESIDE COMPANY

PHOENIX, ARIZ.

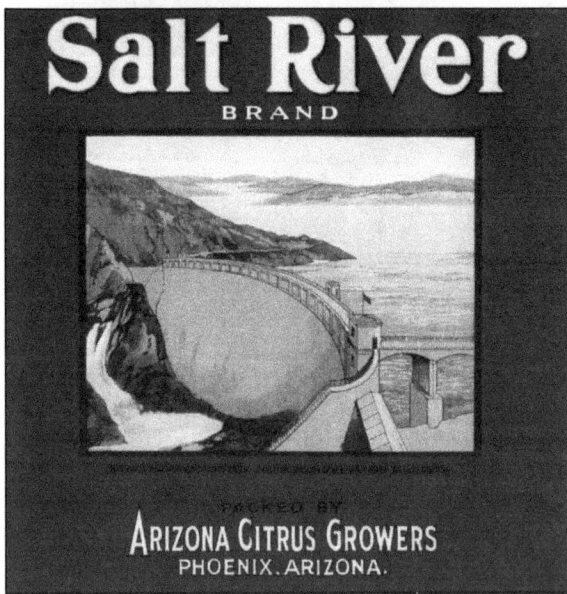

Salt River
BRAND

PACKED BY
ARIZONA CITRUS GROWERS
PHOENIX, ARIZONA.

Agriculture was the economic core of the Salt River Valley, as represented by one of its most important cash crops, citrus. Above around 1893, from perhaps the first orchard planted, the Ingleside crate label represented the success of Arizona Canal builder William J. Murphy and others. Dating from the 1920s, the Salt River Brand (right) illustrates the continued success and growth of Murphy and his associates, which was made possible by the Salt River Project's Roosevelt Dam. Completed by the Bureau of Reclamation in 1911, it was the first of a series of dams along the Salt River and the impetus for projects conceived along the Verde River. (Both, courtesy of author's collection.)

Photographed above on his homestead in Scottsdale in 1900 are retired U.S. Army chaplain Winfield Scott, his wife, Helen, and their mule Old Maud. Scott promoted agriculture in the Salt River Valley and encouraged the introduction of citrus. He and William J. Murphy (right) established two of the area's first citrus orchards. Scott's was a 7-acre plot with 50 trees near the town that bears his name, Scottsdale. One and a half miles away, Murphy planted a 16-acre orchard with several hundred trees in what would become the Arcadia neighborhood of Phoenix. The first harvest occurred in 1891, producing both choice citrus and a showcase to promote land sales for Murphy and his investors. (Both, courtesy of the Scottsdale Public Library)

Looking more like a postcard from Florida's Everglades than the Arizona desert, this undated photograph shows a section of William J. Murphy's Arizona Canal near the Ingleside Orchard in Phoenix. It stretched some 47 miles, beginning just below the confluence of the Salt and Verde Rivers. Water was diverted in the canal by the Arizona Dam. It ended in the town of Peoria, which Murphy helped to establish. As one of the most successful irrigation projects up to that time, the Arizona Canal was one of the most significant developments in the history of Phoenix and Central Arizona. It was acquired by the federal government in 1906 and operated by the Salt River Project since 1907. (Courtesy of LOC, HAER ARIZ, 7-PHEN.V, 1-11.)

Above, nurtured by Arizona Canal water is a newly planted Ingleside orchard section around 1910. Older trees and Camelback Mountain stand in the background. For all its success, the Arizona Canal also underscored limitations of Phoenix's early canal system. Land north of the gravity-fed canal could not receive Salt River water, and without a storage dam, farmers were at the mercy of drought cycles. (Courtesy of the Scottsdale Public Library.)

Prosper P. Parker was inspired by the Arizona Canal. To the left, he is seen around 1890. With Augustus C. Sheldon and Samuel C. Symonds, he planned to bring Verde River water to the north of the Arizona Canal not only with the help of a canal but also with the help of a storage reservoir. The Rio Verde Canal Company formed in 1891 to bring plans to fruition. (Courtesy of Arizona State Library, Archives and Public Records, History and Archives Division, Phoenix, 97-7740.)

The Horseshoe Dam site, photographed in 1943, is about 12 miles north of the Bartlett Dam site. It was the choice of the Rio Verde Canal Company for its proposed 205,000-acre-foot reservoir featuring a 70-foot-high dam with a crest length of 1,200 feet. Rather than the site of the first modern dam on the Verde, it would be the location of the last. (Courtesy of LOC, HAER ARIZ, 13-CACR.V, 1-32.)

Before going bankrupt in 1898, the Rio Verde Canal Company did manage to dig this 730-foot-long diversion tunnel, photographed in 1944 after being cleared, near the Horseshoe site. It was originally intended to divert Verde River water into the company's canal, of which only a few miles were completed. (Courtesy of USBR.)

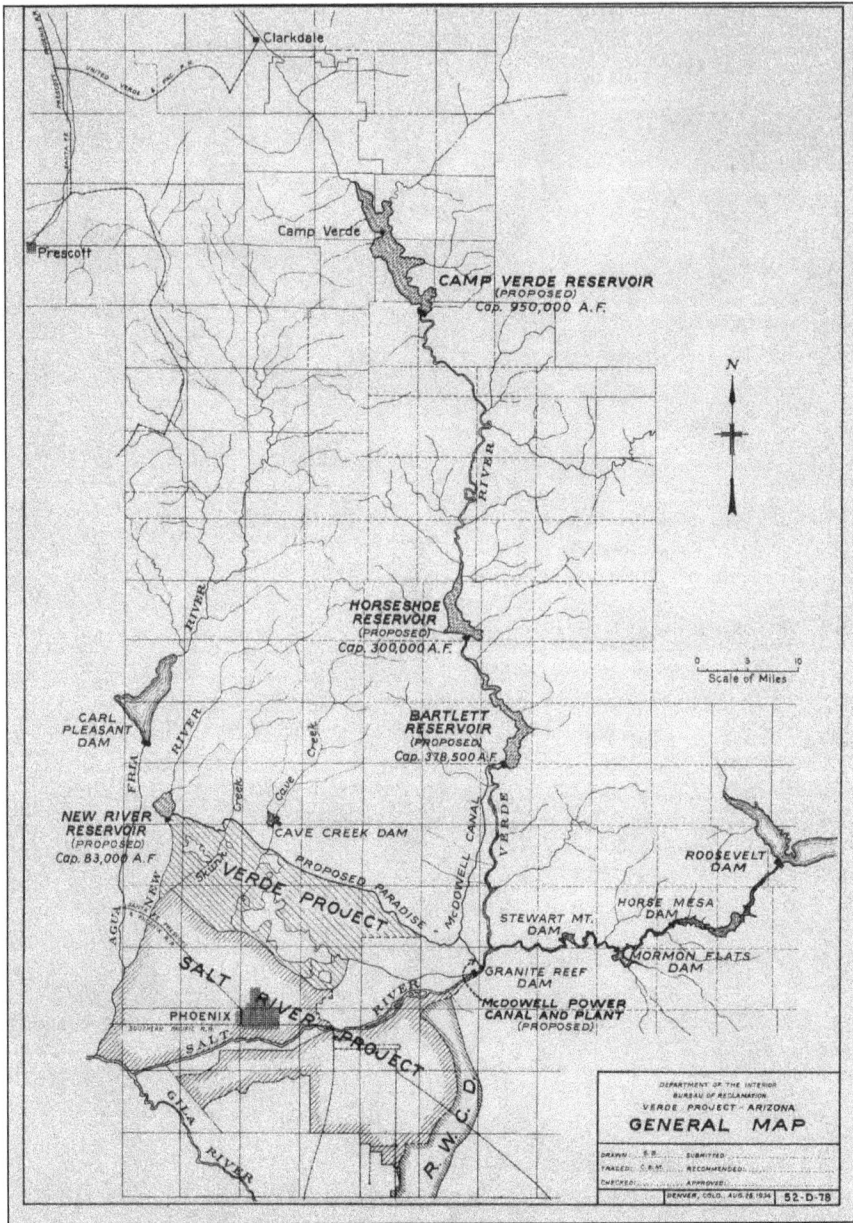

This map from 1934 shows the proposed reclamation projects on the Verde as well the existing dams along the Salt River and its tributaries. The defunct Rio Verde Canal Company had been replaced by a new group, the Paradise Valley Water User's Association, in 1914, the same year the Salt River Project indicated its intensions to build on the Verde. In 1923, after several name changes, the association was known as the Verde River Irrigation and Power District (VRIPD). The two fought bitterly over rights on the river. As late as 1933, VRIPD was going forward with its plans with blessings of the Department of Interior. However, an 11th-hour decision in 1934 determined that the Salt River Project was most suited to operate on the Verde. Many of VRIPD's plans are indicated on this map, which include the following: a huge reservoir near Camp Verde; the Paradise McDowell Canal, which would have irrigated Paradise Valley; a power plant on the canal; and the Bartlett Dam site. Of these, only the Bartlett Dam became a reality. (Courtesy of USBR.)

Pictured some time during the late 1920s in the field and looking like he just stepped out of Steinbeck's *Grapes of Wrath* is Verde River Irrigation and Power District secretary William Bartlett. Although rivals, the Salt River Project kept Bartlett's name for the dam and reservoir for reasons that are unclear. It was William Bartlett who determined the best site for a dam on the Verde River. Seen below is a topographical map from 1936 showing the site where the dam would straddle the gap between two granite cliffs along a north/south axis (top of map). Also depicted is the lake at the bottom of map. The Verde River Irrigation Power District lingered on into the 1950s, with Bartlett staying on as secretary. Their continued pursuits of water rights on Verde River were fruitless. (Both, courtesy of USBR.)

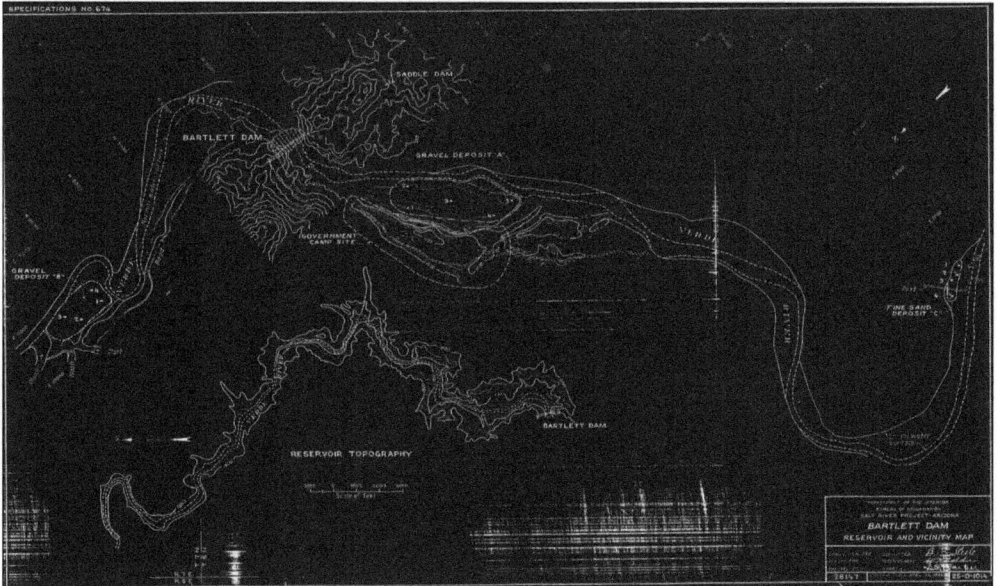

Here are two views of the Bartlett site in 1936. Pictured above are the granite cliffs that would be the dam's abutments. Pictured below are the Salt River Valley Water User's Association members meeting at the site. With them is a scale model of the Bartlett Dam. Bartlett is a multiple-arch dam inspired by the work of Fred Noetzli and designed by Salt River Project consulting engineer Raymond Hill and Bureau of Reclamation engineer Edward C. Koppen. This design offered great strength and cost savings, requiring much less concrete than traditional dam structures. The Bartlett Dam includes 11 arches that are arranged in slight upstream curves spanning nine hollow buttresses. The arches extend from a banked spillway on the north abutment toward the south abutment. As originally built, it was 800 feet wide and 287 feet high. (Both, courtesy of RTM.)

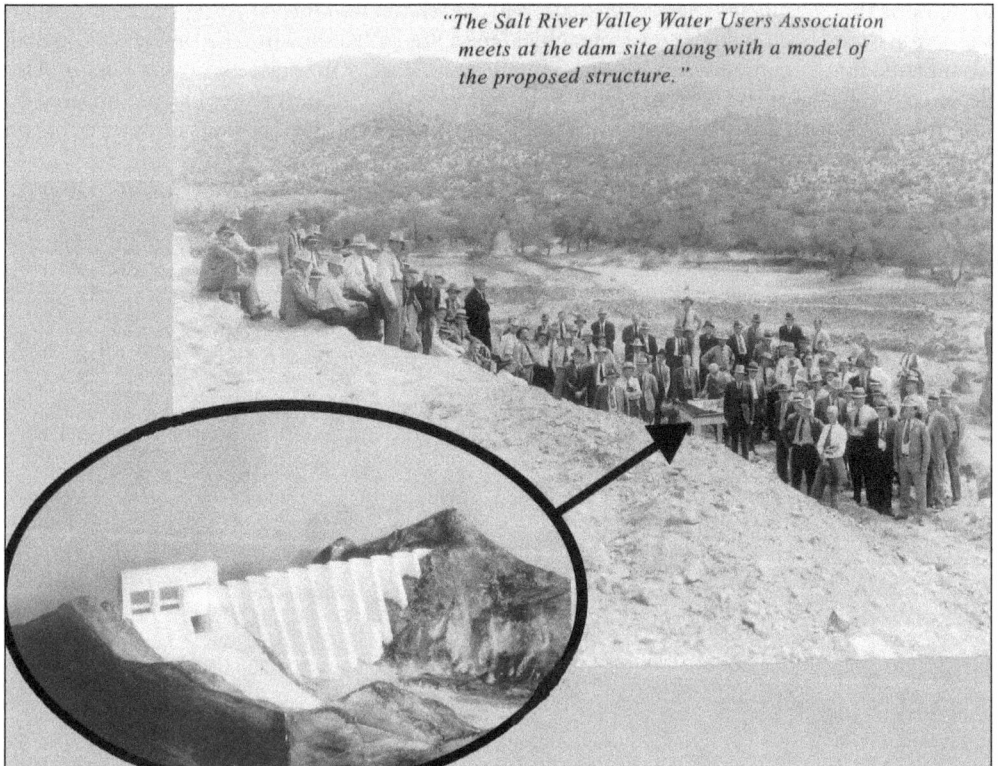

"The Salt River Valley Water Users Association meets at the dam site along with a model of the proposed structure."

Ironically, the multiple-arch design chosen by the Salt River Project was similar to one used by their rival, the Maricopa Water District, for the Waddell Dam on the Agua Fria River (Lake Pleasant). The Waddell Dam, shown above in 1927 near completion, did not become operational until 1936, mainly due to intense challenges from SRP over purported safety concerns with multiple-arch designs. (Courtesy of the McClintock Collection, Arizona Room, PPL.)

One of the controversies surrounding the Waddell Dam involved the formation of cracks on its hollow buttress walls, which can be seen in this photograph taken inside the dam buttress in 1988. (Courtesy of LOC, HAER ARIZ, 7-PHEN.V, 55-66.)

The blasting of the south abutment at the Bartlett Dam was photographed in 1936. The construction contract was awarded to Hilp and Macco Corporation of Clearwater, California. Their bid for the project was $2,228,272. After site preparation, they were given 1,000 days to complete the dam by May 9, 1939. (Courtesy of RTM.)

Before pouring cement, the site needed to be prepared. This involved excavation into solid rock and grouting of fissures. It took the equivalent of 24,220 sacks of cement to fill nearly 34,000 linear feet of grout holes. This photograph shows excavation on the south abutment in 1938. A total of 482,000 cubic yards of rock and soil were removed from the cliffs and riverbed below. (Courtesy of USBR.)

Though it was not clearly determined to what degree, if any, the Waddell Dam cracks affected dam safety on Lake Pleasant, it was determined that they were caused by heat dissipating from the cooling concrete as it set. The solution for this problem was applied on the Bartlett Dam. This 1938 photograph shows the saw-tooth-shaped contraction joints incorporated into Bartlett's buttresses. These spaces were filled when the concrete cooled. In addition, two 18-inch stiffener walls were added to the buttress sides every 41 feet in height. This contributed to an almost flawless concrete structure. Though relatively inexpensive to build, multiple-arch dams required many highly skilled workers to construct. However, Bartlett was raised during the Great Depression, and labor shortages were not a factor. At the time it was completed, the Bartlett Dam was the tallest multiple-arch dam in the world. (Courtesy of USBR.)

Another solution to pouring concrete in the blistering hot desert climate can be seen in these 1938 photographs of spillway construction, above, and buttress work, below. The wet burlap covering the spillway retaining wall was kept wet with a fog spray. This innovation cooled the concrete as it set, allowing for pouring in otherwise prohibitively hot temperatures that reached above 95 degrees Fahrenheit. The fog spray, seen in action on buttress number three, was also used on the contraction joints, shortening the wait to fill them from 90 days down to about 18 days. (Both, courtesy of USBR.)

These two photographs, above in 1938 looking downstream from the north abutment of the dam and below in 1939 looking upstream, show the aggregate processing plant. It was located next to the government camp just as the river bends south again and comes out of the saddle where the dam was built on the west bank. It supplied sand, gravel, and rock for the dam and its cement processing plant. Raw materials were excavated from the river bottom using a dragline, dropped into trucks, and dumped into a hopper. A conveyor belt then fed the mixed materials into a screening device called a "grizzly." (Both, courtesy of USBR.)

After being sized in the grizzly, sorted material was washed and piled over steel tunnels, where trucks, such as the one photographed above in 1939, received loads without the need of a power shovel. (Courtesy of USBR.)

Closest to the dam site was the cement processing plant, left, and cableway, which is extending across this 1939 photograph. The plant could store up to 2,000 barrels of cement, with another 1,000 on hand at the railway in Phoenix. Prepared cement was either hauled by crane in 2-yard buckets or ferried up the cableway to hoppers for "pumpcrete" guns on the dam. (Courtesy of USBR.)

This 1938 photograph is of cement being poured over the spillway drains using 2-yard buckets, which were hoisted by a giant crane. The concrete used over the drains was a porous mix that allowed water to infiltrate into the drains, helping reduce peak runoff velocity and volume on the spillway. (Courtesy of USBR.)

Workers on the south abutment in 1938 fill "buggies" from the pumpcrete hopper on the platform and pour it down chutes, vibrating it during placement. At higher elevations, it was poured from a bucket on a cableway. This backbreaking work allowed a pouring rate of 50 cubic yards per hour. Concrete batches were tested no less than three times from mixing to placement. (Courtesy of USBR.)

This is a view of the downstream faces of buttresses 6 through 10 taken from the top of the cement processing plant in 1938. The light-colored vertical lines on the buttress sides are where the saw-toothed contraction joints have been filled. The bridge-like structure extending above the dam is the pumpcrete trestle. The buttresses were not placed parallel to each other. Instead, they fan out slightly on the upstream side at two and a half degrees like fingers spread apart on a hand. The buttresses, upon which the arches would rest, were designed to absorb the weight of water transferred across the arches, sending it to the foundation. Cement thickness of the buttresses and arches varied from 7 feet at the base to about 2 1/3 feet at the dam crest. The abutments were massive concrete crowns poured over the angled cliff tops, giving them a more or less vertical side for the first and last arches to anchor to. (Courtesy of USBR.)

An arch is placed on a buttress in 1938. Two cement forms were used to build arches, one for the inside of the arch and another for outside. The inside was poured first, fog sprayed, and allowed to set. Then steel reinforcement was added, the outside form was placed, and cement was poured. Arches were formed this way from the bottom up in sections called "arch lifts," each taking about eight days to complete. (Courtesy of USBR.)

The arches lean across the buttresses at nearly 45-degree angles. Taken from the inside of one of the arches, this 1939 photograph is of the downstream side. The arches formed complete half-circles and were cylinder-shaped inside and cone-shaped outside. Over 6.7 million pounds of reinforced steel were used in the arches and buttresses at the Bartlett Dam. (Courtesy of Bartlett Lake Mardeina.)

73

In addition to the desert heat, the Bartlett project also had several major floods to contend with. The first one occurred on February 7, 1937, peaking at 62,500 cubic feet per second, the greatest ever recorded up to that date. Normal river flow is about 500 cubic feet per second. This caused considerable difficulty in excavating buttress and arch foundations and destroyed the cofferdams that kept the work site dry. Progress was slow in 1936 and 1937 because of the flooding. The first cement wasn't poured until February 5, 1937. By the year's end, the dam was only about 38 percent complete. This photograph, dated July 30, 1938, not only shows significant improvement in progress in 1938 but also shows a dramatic contrast to the image on page 75. Here the Verde River is at more or less normal flow. (Courtesy of USBR.)

At about the same angle as the photograph on page 74, the flood above was photographed on September 4, 1938, from the south abutment with water flowing downstream from the right to the left. Note the crescent-shaped waves on the right that are breaking just before reaching the buttresses. These are the partially completed dam arches, which have progressed in height from the previous photograph. Here, they are completely submerged. Flow for this particular flood was estimated at 77,000 cubic feet per second. The largest flood occurred earlier on March 4, 1938, coming in at a whopping 108,000 cubic feet per second. Amazingly, little damage was sustained, as construction was far enough along to survive these episodes. If the Salt River Project still required proof of the strength and stability of a properly constructed multiple-arch dam, Bartlett's survivability in the face of these torrents likely satisfied them. (Courtesy of USBR.)

Images of any dam, however well built, overtopping are disconcerting to the public as an unacceptable hazard. Bartlett, like most dams, includes a spillway that was calculated to handle the maximum anticipated stream flow. As originally built, the spillway was designed to handle a flow of 175,000 cubic feet per second at an elevation of 1,798 feet. The plans from 1936 (above) placed the spillway on the north abutment. Below, in 1938, water would be released from three massive Stoney gates. Water is directed along a banked concrete channel, where it discharges into the river off an elevated ramp. (Both, courtesy of USBR.)

A Stoney gate is a crest gate that moves along a series of rollers traveling vertically in grooves in masonry piers, independently of the gate and piers. Bartlett's gates are gargantuan 50-by-50-foot steel behemoths weighing 200 tons each. The worker in this 1939 photograph is reduced to Lilliputian proportions standing next to the roller system, which looks like a massive bicycle chain. (Courtesy of USBR.)

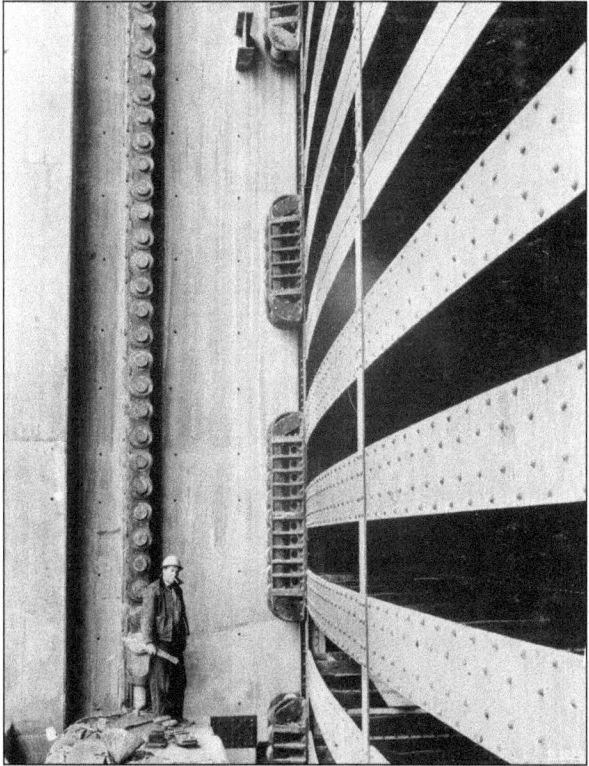

The room at the top of the gates, seen here in 1989, houses the motors and hoists used to raise and lower the gates. Each of the three 7.5-horsepower motors is capable of raising the gate at 4 inches per minute. (Courtesy of USBR.)

In addition to the spillway, two other mechanisms control water. In this 1938 photograph, which was taken between buttresses eight and nine, the three square openings at bottom house the 6-by-7-foot outlets for eclectically operated slide gates used to regulate river flow. In the small building on the right, the circular openings house two 60-inch needle valves for irrigation releases. (Courtesy of USBR.)

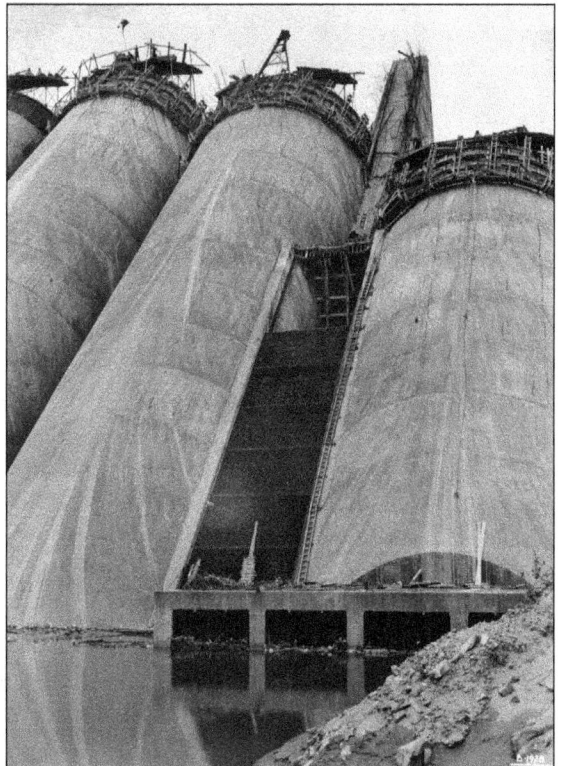

This 1939 view from the upstream side displays the openings for the river-outlet works and needle valves between buttresses eight and nine. They are covered with grates called "trash racks," which were designed to keep reservoir debris out of the inlets. The combined output of these works is 2,600 cubic feet per second. (Courtesy of USBR.)

The 60-inch needle valves, originally installed in 1939, used to control irrigation flows from the Bartlett Dam to SRP customers downstream. The top view is from inside the small building in front of buttress nine. The upper outlet, positioned 40 feet above the river, has been removed for maintenance. Note that the operators are standing above the lower valve. In 1984, air became trapped in the operating chamber of the lower valve, causing a rupture that blew out the almost-2-inch-thick steel valve casing, killing the operator. Below, the lower valve is shown discharging water with tremendous force 28 feet above the desert riverbed. (Both, courtesy of USBR.)

Symbolizing the power politics of water in Arizona, the Bartlett Dam is visited by dignitaries as it rises out of the Verde River in October 1938. Pictured are, from left to right, Harold J. Lawson, Salt River Project's chief engineer, who argued that the Salt River Project was the only entity that could successfully develop the Verde; Allen Mattison of the U.S. Bureau of Reclamation; Sen. Carl Hayden, who, like other Arizona politicians, threw his support to the Salt River Project after the *Debler Report* in 1934; Lin B. Orme, Salt River Valley Water User's Association's president, who was the main force behind SRP's campaign to win the Verde from the Verde River Irrigation and Power District; Paul Roca, secretary to Senator Hayden; and Joseph A. Fraps, civil engineer for the State of Arizona, who investigated the cracks on the Waddell Dam. (Courtesy of USBR.)

By May 1939, the Bartlett Dam was finished, ahead of schedule and below cost. Bartlett Lake extends 12 miles north of the dam, covers 2,815 surface acres, and can hold 178,186 acre-feet of water before the spillway have to be opened. It was calculated for a probable maximum flood (PMF) around 175,000 cubic feet per second, or so it was thought. (Courtesy of LOC, LC-DIG-ppmsca-17399.)

On Valentine's Day, 1980, the Salt River system experienced a near-catastrophic flood. Hydrologists discovered the true PMF at the Bartlett Dam was an astonishing 562,000 cubic feet per second, which was more than double what had been previously thought. Above, the U.S. Bureau of Reclamation's Hydraulic Laboratory studies spillway modifications on a scale model of the Bartlett Dam in the early 1990s. (Courtesy of USBR.)

Between 1994 and 1996, modifications were made to the Bartlett Dam to prevent overtopping. It was decided to raise the dam height 21.5 feet from 287 to 308.5 feet. Shown are two views of the arches being raised. On the upstream side (above), a huge crane on the south abutment helps place forms on one side of the dam, while a barge with a floating crane (below) works from the north abutment. (Both, courtesy of RTM.)

Studies conducted by the Bureau of Reclamation showed the original spillway could be modified to handle 287,500 cubic feet per second at a higher elevation of 1,821 feet. That was still not enough. A new spillway was created in a cove about a quarter-mile south of the dam. Above, the new spillway that is under construction in 1996 is a "fused plug" spillway. As water levels in the lake exceed 1,821 feet, water will begin to cascade down the mountainside on the other side of the cove, seen below in the 2009. The spillway is filled with rubble that will wear away, like a fuse as water flows, allowing for some control over the flow rate. The auxiliary spillway adds an additional 261,700-cubic-foot-per-second capacity. (Above, courtesy of RTM; below, courtesy of author's collection, Tonto National Forest, permit CVC581.)

The Bartlett Dam is photographed in 1996, as modifications were nearly complete. The new auxiliary spillway can be seen wedged into the mountain on the right. At flood stage, the lake can now hold up 3,500 surface acres with a capacity of 249,686 acre-feet. The crest length of the modified dam is 1,130 feet. The entire infrastructure from the dam's 1936–1939 construction

has been removed, leaving a blank space on the river (bottom left), which is where the aggregate processing plant used to be. The Bartlett Dam has no hydroelectric generating capacity at this time. (Courtesy of RTM.)

With the Mazatzal Mountains looming in the background, boaters enjoy a day on Bartlett Lake in this 2009 photograph. The dam has created a lake almost 200 feet deep and nearly a mile wide in places. A body of water such as this had not been seen in the Verde Valley for millions of years, when the water flowed in the opposite direction. Most are probably unaware of the long history of people's struggle to control the water, both among each other and against nature, which brought Bartlett Lake into being. (Courtesy of author's collection, Tonto National Forest, permit CVC581.)

Six

ANOTHER WAR, ANOTHER DAM

On June 19, 1915, a delegation from Arizona breaks two bottles across the prow of the USS *Arizona*, the Navy's newest battleship. Inside one of those bottles is the first water to touch her hull, water from Arizona's Salt River. The sinking of this ship at Pearl Harbor 26 years later would cause construction of the Salt River Project's Horseshoe Dam. (Photograph, United States Navy, courtesy of NavSource, http://navsource.org/.)

Phelps Dodge Company's open-pit copper mine in Morenci, Arizona, was photographed in 1942. During the late 1920s and 1930s, Phelps Dodge developed methods of extracting copper from low-grade "clay orebody." To be profitable, the ore had to be mined in vast quantities. Instead of traditional underground mineshafts, huge pits were excavated, allowing ore to be put into railcars and transported to the nearby processing plant. (Courtesy of LOC, LC-USW3- 027813-E.)

Shown here is a wartime image of a copper ore concentration at Morenci in 1942. Ore is crushed, mixed with water, and ground to a pulp. The pulp is placed in flotation cells and chemicals are added, while constant agitation lifts copper into a froth that is skimmed off for further refining. The process required 175 gallons of water for every ton of ore. (Courtesy of LOC, LC-USE6- D-010041.)

Ordered by the War Production Board to increase copper production by 80 percent, Phelps Dodge needed a major new water source. They chose the Black River in the White Mountains of eastern Arizona. Like the Verde, the Black River is a major tributary of the Salt River. The Salt River Project considered it part of their system. By 1943, Phelps Dodge worked out an agreement with SRP for a water exchange. To compensate SRP for water taken from the Black River, Phelps Dodge would construct a storage dam at the Horseshoe site on the Verde to store runoff that would otherwise be lost. These photographs from 1945 show the Black River diversion works. The pump house and diversion dam are pictured above, and below, the gauging station is at the end of the pipeline. (Both, reproduced with permission of Freeport-McMoRan Corporation.)

The Horseshoe site, so named because of the horseshoe-shaped bend in the Verde River, was photographed in August 1944. The old Horseshoe Bridge is still in place as the site is prepared for the Salt River Project's last dam on the Verde. One of the first tasks involved is cleaning and sluicing the dam abutments. In this photograph, a high-pressure hose washes away loose rock and debris from the west abutment, which is where the spillway would be located. The process was not unlike hydraulic mining techniques used by gold mining operations throughout the West beginning in the 19th century. Compare the images on this page and page 91 with the relatively untouched west abutment shown on page 45. It shows the radical transformation of the environment that was necessary to dam the river. (Courtesy of USBR.)

Note that all that remains of the Horseshoe Bridge in October 1944 are the two suspension towers on either shore. Association shareholders voted twice before approving the project and, in doing so, dashed any hopes the Verde River Irrigation and Power District had for gaining a share of the river. To this day, Paradise Valley receives no Verde water. The design selected for the site was unlike any other dam built for the Salt River Project. Unlike the Bartlett Dam, which relies on its multiple arches to transfer the force exerted by the water's weight to the dam's foundation and abutments, an earth-fill dam is an embankment-type dam that forms a massive barrier that is pressed down upon by the force of impounded water. The earth-fill dam is one of the oldest types of dams, dating back to 4000 BC. Nevertheless, it was a source of alarm for many in the Salt River Valley. (Courtesy of the U.S. Bureau of Reclamation.)

The earth-filled Walnut Grove Dam was photographed on the Hassayampa River in Yavapai County around 1888. The dam, completed in 1887, was poorly constructed, and after heavy rainfall in 1890, it failed, sending a 40-foot wall of water toward Wickenburg that killed about 100 people. The more recent failure of the earthen Lyman Dam on the Little Colorado River killed two and caused hundreds of thousands of dollars in property damage. These and other dam failures were on people's minds when the Horseshoe design was suggested. The two main causes of failure in earthen dams are overtopping, which causes the face of the dam to erode away quickly, and seepage, in which the dam's integrity is compromised by water seeping inside the dam, weakening the structure. (Courtesy of the McClintock Collection, Arizona Room, PPL.)

The solutions to the concerns over the Horseshoe Dam design are shown in these plans from 1944. To the right is an overhead view of the dam showing a very large spillway on the west abutment as well as the girth of the dam itself. Described as a "zoned" earthen structure, the dam features a compacted core that is impervious to water and is surrounded by a sand and gravel intermediate zone that binds to a thick outer layer of rock fill. The result was a massive structure, seen in this 1944 cross section—650 feet wide at its base, 40 feet wide at the crest, 1,500 feet long, and 144 feet tall. (Courtesy of USBR.)

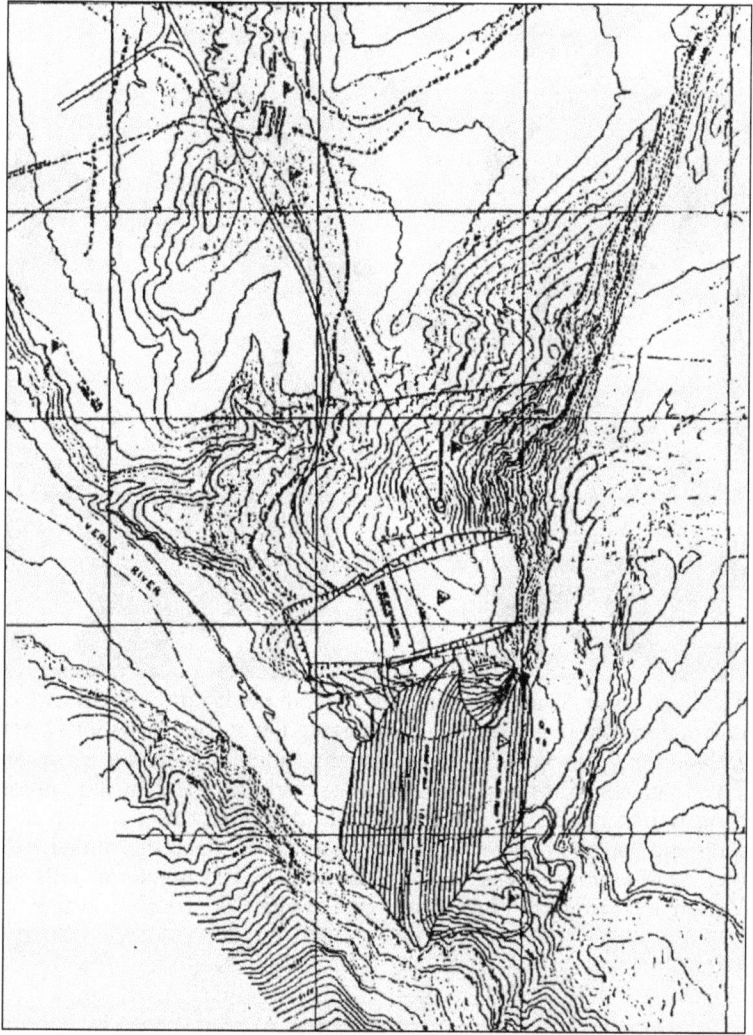

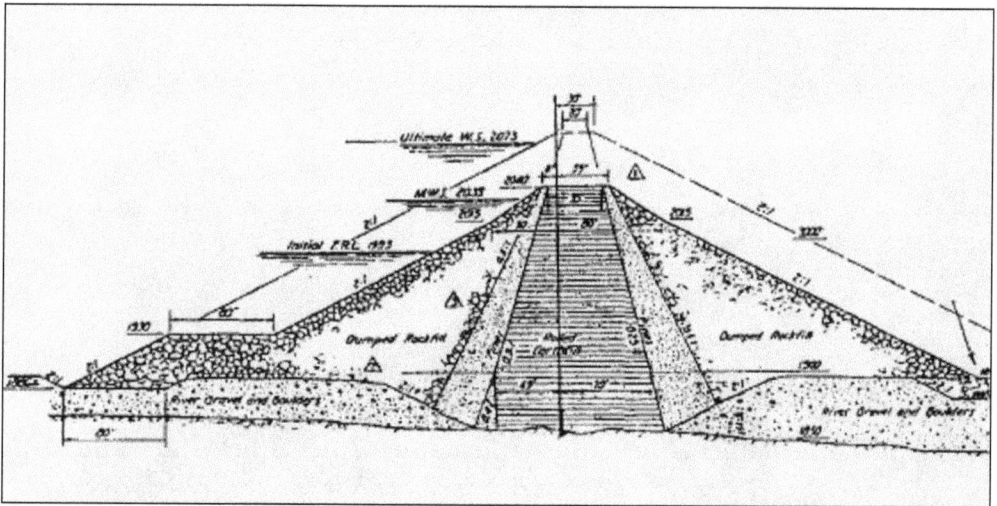

The Dixon and Arundel Group of Baltimore tendered the winning bid of $1,656,349. The first task undertaken in preparation for construction of the Horseshoe Dam was to clear and line the 730-foot-long Rio Verde Canal Company tunnel, which was excavated in 1896 (see page 61). This would serve as the main river outlet works for the dam. The 14-foot-wide tunnel was large enough to fit a bulldozer inside to plow the debris out the end. Then crews in 1944 smoothed the walls and pumped concrete into forms (above) to shape the tunnel walls and floor. The ceiling (below) was sprayed on using gunite to complete the arch form. This work was subcontracted to Vinson and Pringle of Phoenix at a cost of $164,000. (Both, courtesy of USBR.)

Above, at the upstream entrance to the tunnel in 1944, an outlet tube is fitted at the location of the base where the outlet tower will be constructed. Below, also in 1944, on the downstream side of the tunnel is the river outlet. The water from the tunnel, once intended to irrigate the Paradise Valley area of Phoenix (via a canal), now empties into Bartlett Lake at its northern most end. Although constructed similarly to hydroelectric inlets for power-generating dams, the Horseshoe Dam, like the Bartlett Dam, has no such capabilities. (Both, courtesy of USBR.)

In 1944, workers began to prepare the dam foundation. After clearing away loose dirt and rock with pressurized hoses, they discovered a subterranean stream-flow seeping out of the bedrock at 500 gallons per minute. This hampered construction efforts and threatened dam stability. The solution involved placing two 14-foot-high concrete-core walls along the upstream and downstream edges and along the dam's core section, concentrating water so it could be pumped out. A series of 1.5-inch holes 25 to 100 feet deep were drilled into the foundation and sealed with pressurized cement, checking the seepage. These photographs from 1944 show the first cut in the west abutment for the core wall (left) and one of the core walls (below) that extends across the foundation as workers prepare bedrock. (Both, courtesy of USBR.)

To ensure against further seepage problems, a 100-foot-long, 10-foot-high concrete cutoff wall was constructed over an 8-foot-thick concrete slab, which was placed at the deepest point of the gorge. Seen above in December 1944, the first section of the rolled-earth core is placed over the slab and compacted down to form the Horseshoe Dam's mostly impervious center. (Courtesy of USBR.)

As seen here in September 1945, the Horseshoe Dam rises looking toward the west abutment. The core of the dam is being covered with the sand and gravel "transition zone" (center), while the rock fill (right) is being added to the outermost zone. (Courtesy of USBR.)

Upon completion of the outlet tunnel, which is seen here in February 1945, construction of the outlet tower begins. The elbow-shaped tube has been encased in a concrete slab that will serve as the tower base. (Courtesy of USBR.)

The tower was fabricated on site at the dam's concrete precasting plant. Hundreds of circular segments are being cast and set in this photograph from February 1945. The segments were hauled and stacked over the outlet tunnel opening on the upstream side of the dam using a wooden framework. (Courtesy of USBR.)

In April 1945, the tower, located at the northwest corner of the dam, is nearly 10 stories tall. The segments were assembled into circles and held together by 6-foot struts that were cast in forms, as seen at the top of the tower. The openings in the tower were covered with steel trash racks, which keep large debris out of the tunnel. (Courtesy of USBR.)

When completed in 1945, the tower was topped with a service bridge (left) and a gauging station. The gauging station controls water release with a 9-foot-diameter steel plug at the tower base that is raised and lowered using a 12-ton hoist and a 92,500-pound counterweight. (Courtesy of USBR.)

With the Walnut Grove disaster on people's minds, as well as the Verde's propensity to flood, an overtopping of the Horseshoe Dam was a fearful prospect. Consequently, great care was taken designing the dam's spillway. The crest elevation of Horseshoe is 2,040 feet above sea level. The spillway elevation is at 1,993 feet—a full 47-foot difference. The 320-foot-wide concrete structure can handle a 40-foot-high wall of water spilling over its lip at 250,000 cubic feet per second. These are photographs of the spillway's construction in 1945. Above, the left side of the spillway sidewall is being poured into forms containing steel reinforcement. Pictured below is spillway paving looking toward the east abutment. Prior to this, the underlying bedrock supporting the spillway was pressure grouted using more than 3,000 bags of cement. (Both, courtesy of USBR.)

Two views of the downstream side of the Horseshoe Dam spillway are pictured around 1945. Designed to be functional rather than aesthetic, the structure nevertheless possesses a certain elegance of form. Above, the channel cut into the bedrock to divert water flow into the river can be seen. Below is a view of a unique design feature—a lip extends out and creates a 15-foot waterfall over the bedrock channel. The lip supports a roadway that allows access across the spillway to the dam. When water is flowing over the lip, the east side of the dam may still be accessed by the walkway that can be seen under the lip. Chief engineer Raymond Hill defended this somewhat controversial feature, noting that access to the outlet tower needed to be available under all conditions. (Both, courtesy of USBR.)

This 1951 photograph shows an upstream view of the spillway. One of the reasons for the nearly 50-foot difference in height between the dam crest and the spillway elevation was to allow for the addition of tainter gates. The moveable gates were contemplated almost as soon as the dam was completed. They were added to enlarge reservoir capacity. The main reason for this was so the Salt River Valley Water User's Association could retain and control most of the floodwater on the Verde. By doing so, they could preclude other groups, such as the Verde River Irrigation and Power District, from claiming flow in excess of the dam's capacity. The construction of the gates was paid for by the City of Phoenix in exchange for a share of the water that was stored at Horseshoe. This served the Salt River Project, in that the city of Phoenix was within the SRP service area. This ensured that the lower Verde River would be solely under SRP control. (Courtesy of USBR.)

When the Horseshoe Dam was built, it was thought it would be expanded to accommodate water from the Colorado River one day. These plans were soon put aside, and construction of the gates began in 1949. Above are the three 106-foot-long gates on the downstream face of the spillway. They doubled the capacity of Horseshoe Lake from 67,000 to 139,000 acre-feet. (Courtesy of USBR.)

The 33-foot-high steel gates are raised and lowered by lifting motors and counterbalances as seen on the right in this 1951 photograph. The gates allow for river control in addition to the outlet tower and tunnel. They are equipped with automatic sensors and regulate the reservoir level. They can also be controlled manually. (Courtesy USBR.)

The original agreement between the Phelps Dodge Corporation and the Salt River Project called for construction of the dam in exchange for 250,000 acre-feet of Black River water for the Morenci operation. Black River water is no longer used at Morenci, and the Horseshoe Dam and has been used solely to service the Salt River Project since 1951. Ironically, the dam wasn't completed until 1946, several months after the conclusion of World War II. If the Rio Verde Canal Company or any of its successors had prevailed, the history of the lower Verde and the lands in Phoenix, north of the Arizona Canal, would have been written very differently. Had the Salt River Valley Water User's Association not asserted itself in the Verde Valley, it is likely the Bureau of Reclamation would have developed dams and resources there with another entity before the 1930s. The completed Horseshoe Dam looks today much as it did in this *c.* 1951 image—a monument to man's ingenuity and the "water wars" of the desert Southwest. (Courtesy of USBR.)

Seven

CAMP LIFE

Vernon William Davidson and Clara Constance "Connie" (Neagle) Davidson of Barnes County, North Dakota, pose for their wedding portrait in 1934. Vernon practiced farming until the droughts of the Dust Bowl caused the Davidsons to auction their farm in 1935. Vernon found work on the dams being built in Arizona. Together they lived and worked at the Bartlett Camp. (Photograph by Vernon and Connie [Neagle] Davidson, courtesy of V. W. Davidson Collection.)

Photographed looking north around 1938, here is a rare view of the Bartlett camp. The Verde River is in the foreground. The camp was located on a long, narrow bluff above the aggregate processing area. On the downstream (left) side in the shaded area were the "white overflow" and Native American camps. Workers were segregated from each other. It is estimated that about a dozen tents were located in the Native American area, and the Yavapai most likely occupied

this area. Immediately to the right of this was the government section, which housed Bureau of Reclamation engineers, inspectors, and staff. The larger buildings were dorms and a garage. Upstream from there was the contractors area, which housed a school, a hospital, a service station, a recreation hall, and housing and dormitories for the workers. (Photograph by Vernon and Connie [Neagle] Davidson, courtesy of V. W. Davidson Collection.)

The V. W. Davidson family in Phoenix in 1943 includes, from left to right, Connie, who is holding Victor; La Verna; and Daryl, who is being held by his father, Vernon. All the children were born in Phoenix; however, La Verna was with her parents at the Bartlett camp as an infant. She and her mother were among the relatively few females present at the camp. (Photograph by Vernon and Connie [Neagle] Davidson, courtesy of V. W. Davidson Collection.)

An early recollection of Connie's camp life was that there was an abundance of rattlesnakes, which is a recurring theme at the Bartlett Dam. As early as 1934, the Salt River Valley Water User's Association sent two men to camp at the dam site to protect the association's claims on the river. They reported numerous encounters with rattlesnakes, such as the diamondback pictured above. Sometimes they spotted dozens of snakes at a time. (Courtesy of the McDowell Mountain Regional Park)

Here are two views of the Davidsons' trailer from 1938. Upon first arriving at the Bartlett camp, Connie Davidson recalled the following: "This is great, a real home. My husband, his brother, brother-in-law, and friend had cut and leveled a place out of the side of a small hill and built a two-bed frame room across the back of the trailer and one down the open side, which was where we cooked and ate, while overlooking the main camp site that was well established before we got there." Much of the housing consisted of tents. An area was cleared above the east end of camp for a relatively new form of housing, trailer homes. Connie Davidson, like many during the Depression, felt fortunate that her family had work and any place to live. (Both, photograph by Vernon and Connie [Neagle] Davidson, courtesy of V. W. Davidson Collection.)

Here is a view photographed from the east end of the contractor's area in 1938. The rectangular buildings were dormitories. The hospital is located at left between the dormitories. The large building on the right is the mess hall. Connie was glad to have a doctor present, but noted, "The men knew I worried about them, so they very rarely told me of accidents." (Photograph by Vernon and Connie [Neagle] Davidson, courtesy of V. W. Davidson Collection.)

Another view of the east end was photographed from the trailer park, not far from the hospital, in 1938. "My husband and his brothers all worked setting forms and pouring concrete," recalls Connie. Wages were around 50¢ an hour. Work was hazardous indeed. Three men died that year in an avalanche caused by jackhammer vibration. (Photograph by Vernon and Connie [Neagle] Davidson, courtesy of V. W. Davidson Collection.)

A ghostly image of the Bartlett Dam illuminated in the dark was taken in 1938. Work, as well as the hazards of living in the desert, went on 24 hours a day for the women and men. "A regular day for me was cooking, washing dishes, and taking care of baby La Verna. With the three-work shifts, there was always someone to feed a lunch. I also gathered wood for cooking when I had time. The best part [of camp life] was having my family with me, and for part of the time, my mother was there too. The worst was probably the long drive to Phoenix to buy groceries. The campsite flooded in a storm. The cofferdam broke and did some damage to the work site, equipment, and I think some in tent city. Of course, we were up on the hillside above, but the running water could still be seen," said Connie (Neagle) Davidson. (Photograph by Vernon and Connie [Neagle] Davidson, courtesy of V. W. Davidson Collection.)

Connie said, "The road, of course, was just dirt, very rough sometimes but no sharp curves, no mountains, and had been well traveled." This surviving portion of the original Bartlett Dam Road, shown in 2009, underscores Connie's talent for understatement. A 15-mile segment of the road remained a challenge for vehicles until it was first paved in the early 1990s. (Courtesy of author's collection, Tonto National Forest, permit CVC581.)

Even after it was paved, the desert refused to give way to the modern era. This section of the Bartlett Dam Road that crosses Camp Creek frequently washed out during flash floods until the Maricopa County Highway Department hardened it in 2009. (Courtesy of the Maricopa County Department of Transportation.)

Connie stated, "La Verna was there, a baby in her high chair. For her playtime we sat outside in the sun and fresh air with her dog. The dog took very good care of her. No stranger could come near. My 'men' were all very good to play and help take care of her when they had time. As for tent city children, if any, I am sure they had plenty of room to run and play." Here are two views of the Bartlett Dam Accommodation School from around 1939. (Above, courtesy of USBR; below, courtesy of RTM.)

Here are two views of the last dam camp on the Verde River, the Horseshoe Dam camp, photographed in 1944. Pictured above is the Dixon and Arundel Group camp from the east side looking west. Pictured below is an inside view of the camp mess hall revealing the small but accommodating staff who serviced the no-frills venue. The Bartlett camp housed about 200 men plus 15 families, along with about 16 government staff. The Horseshoe camp housed less than 100. The camps featured heating and cooling (in some buildings), running water, electricity, and telephone lines. The Horseshoe camp was last occupied in 1951, which was when the tainter gates were being installed on the spillway. (Both, courtesy of USBR)

With Phoenix 50 rough miles away, dam workers who just wanted to get a drink and unwind were faced with a daunting trek. Sensing a business opportunity, local Johnny Walker established a favorite watering hole along the road through Cave Creek, Arizona. The Cave Creek Corral is seen in these photographs around the time it was established in 1935. It became frequented by camp workers who enjoyed socializing with local cowboys and miners while downing bottles of Arizona's popular local brew A-1 Beer. It quickly became the local hotspot and has been in continuous operation ever since. (Both, courtesy of Harold's Corral, Cave Creek, Arizona.)

Harold Gavin and his wife, Ruth, took over the business in 1950. They developed it into a local institution and renamed it Harold's Corral. It has become a Cave Creek landmark, though it has changed owners several times. It has also expanded exponentially in size over the years. Seen here in 2008, the original one-room frame building (with the sign) has been preserved and incorporated into the ultimate cowboy bar and grill with two bars, a live stage, a dance floor, and seating inside and out for 800. In addition to locals and tourists, it is still frequented by visitors from Bartlett Lake, but the dam workers have been replaced by boaters and fishermen. (Courtesy of Harold's Corral, Cave Creek, Arizona)

After the last dam camp closed, the lakes went largely unimproved and uninhabited. Horseshoe Lake remains that way. However, after a 1987 trip to Bartlett Lake, Eric Church (standing) and his brother Brian (seated) wondered why such a popular boating lake lacked a marina. Luckily, the Forest Service was soliciting bids. They won their permit in 1990 and are pictured here in 2002. (Courtesy of Bartlett Lake Marina, Tonto National Forest, permit CVC581.)

Eric set to work in 1991 constructing the facility himself with a few others helping. It was backbreaking labor. Above in a cove near the dam are stockpiles of floating pontoons (left) and aluminum dock frames (top, right). The frames float on the pontoons, which are anchored into place by huge concrete blocks. (Courtesy of Bartlett Lake Marina, Tonto National Forest, permit CVC581.)

Ironically, Bartlett Lake Marina, floating on nearly 180,000 acre-feet of water, must provide its own water. Every drop of Bartlett Lake is used downstream. The Churches located their own well by hiring a dowser. Here in 1991, family member Don Fernie tries his hand at the ancient art. Their well was located 600 feet underground with this method. (Courtesy of Bartlett Lake Marina, Tonto National Forest, permit CVC581.)

In order to accommodate rising and falling lake levels, the cove in which the marina is located was excavated during a low-water period in 1991. The maximum normal elevation of Bartlett Lake is 1,798 feet. The marina can operate as low as 1,746 feet. Spoil from the lake was used to create a parking lot for the marina. (Courtesy of Bartlett Lake Marina, Tonto National Forest, permit CVC581.)

Bartlett Lake Marina started with only 20 slips. In this photograph from 1997, it extends from shore with 50. Originally, no fuel was available. Later fuel was hauled down to the dock. Now it is pumped in. The floating facility is kept in place during fluctuating water levels by winches that raise and lower the docks and modules. (Courtesy of Bartlett Lake Marina, Tonto National Forest, permit CVC581.)

By 2009, the marina had grown to 190 wet slips sheltering everything from small fishing boats to luxury houseboats. A dry storage facility accommodates another 150 boats. In addition to fuel and bait, there is a convenience store, food, floating lodges, and a boat rental club. Almost anything that skiers, fishermen, and boaters need is now available for visitors. (Courtesy of Bartlett Lake Marina, Tonto National Forest, permit CVC581.)

Connie Davidson, now Connie Lightfoot, is pictured in 2008 at age 96. In 1939, Vernon went to work on the Parker Dam on the Colorado River where, in 1942, a scaffold he was working on collapsed, landing him in wet cement with an injured back. He recovered, living until 1969. Connie lives in Glendale, blushing at the suggestion she is an Arizona pioneer and modestly pointing out "that beating rattlesnakes with a broom was just part of daily life at the dams." Nevertheless she is pioneer in every sense of the word, joining the men and women who came west in search of a better life and, in so doing, built a better life for others. When asked what she wants people to remember about building the dams, she said, "I guess the most important thing was though we may have not always had the best of everything; life is hard work. But that was where my husband worked and my place was with him regardless of where; happy and thankful for what we had." (Photograph by Vernon and Connie [Neagle] Davidson, courtesy of V. W. Davidson Collection.)

Eight

THE DAMS THAT
NEVER WERE

Here is an artist's rendition from the early 1970s of the proposed Orme Dam at the confluence of the Salt and Verde Rivers. It would have created a lake extending east along the Salt River from Granite Reef Diversion Dam to Stewart Mountain Dam and north along the Verde from the Salt River to the Bartlett Dam. It would have inundated the Fort McDowell Yavapai Nation. (Courtesy of USBR.)

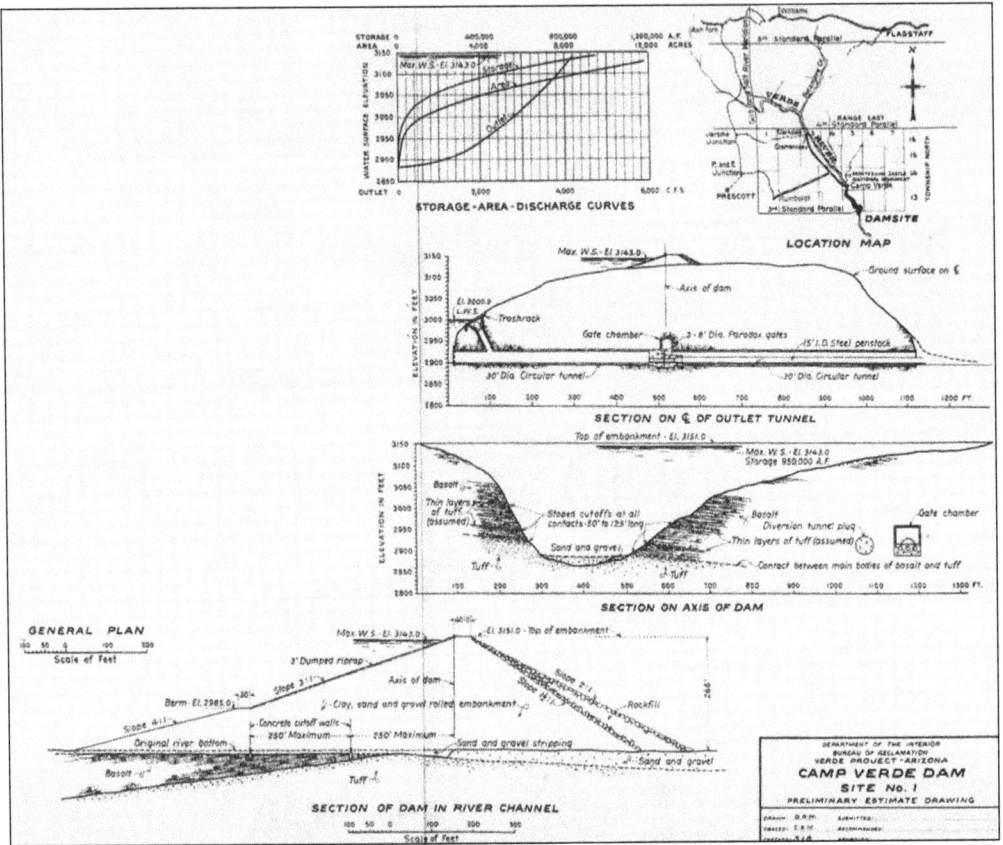

Above from 1934 are plans for a massive earth-filled dam that the Bureau of Reclamation would have constructed for the Verde River Irrigation and Power District had they prevailed in their fight with the Salt River Valley Water User's Association. Below is a photograph of the proposed dam site near Camp Verde (see map, page 62), which would have been well north of the sites of the Bartlett and Horseshoe Dams. The dam would have been 266 feet high with a crest length of 1,300 feet, impounding a gigantic 950,000-acre-foot reservoir—more than five times the size of Bartlett Lake. (Both, courtesy of USBR.)

Another Verde River dam that existed purely in the minds of artists and engineers was the Cliff Dam. It was to be located on Bartlett Lake, below Horseshoe Dam, and the resulting impound would have covered Horseshoe Dam. "Plan Six" was actually part of an alternative to the Orme Dam, which was defeated in 1981. This plan shifted Central Arizona Project's Colorado River water storage from the Verde to the Agua Fria River, behind the New Waddell Dam at Lake Pleasant. Cliff Dam was part of the flood control planning of Plan Six, rather than Central Arizona Project storage, and this portion was dropped after opposition from environmentalists, who were chiefly concerned with bald eagle populations. It was the catalyst behind the modification of the Bartlett Dam during the 1990s. Ironically, the enlarged Lake Pleasant that resulted from the New Waddell Dam is the site of the most successful bald eagle enclosure in Arizona. (Courtesy of USBR.)

Of all the "dams that never were," the Orme Dam was the most likely to have been constructed. The defeat of the Orme Dam came at the hands of the people who, more often than not, had found themselves on the losing end of Verde River history, the Yavapai. The Central Arizona Project, approved by Congress in 1968, called for pumping Colorado River water over 300 miles upstream to the new dam. The new reservoir would have inundated nearly 17,000 acres of the Yavapai land, including homes, burial grounds, and archaeological areas. Though offered $33.5 million and 2,500 acres of land as compensation, they rejected the offer and began a 10-year struggle to save their homeland. Above in 1966 are two tribal members who would take up the fight, Bootha Brown (far left) and tribal chairman Dr. Clinton Patea (next to him). With public opinion squarely on their side, Reagan administration Interior Secretary James Watt announced cancellation of the Orme Dam in November 1981. (Courtesy of the Department of Interior.)

There were several agencies that provided law and order along the dams that were built. The Bartlett Lake Emergency Aid Station, photographed in 2009, was completed during the early 1990s and is located near the main boat launching area. The Maricopa County Sheriff's Office, Lake Patrol Division is tasked with law enforcement and public safety within Tonto National Forest recreation areas, which includes several of the lakes located there. Deputies who are assigned to the Lake Patrol Division are specially certified emergency medical technicians, or paramedics. Also at the ready are a dive team and a detective unit that investigates crimes and accidents on the water and around the lakes. They have a variety of vehicles at their disposal, including patrol boats, such as the one seen on the left. The Lake Patrol Division emphasizes education and common sense on the water. Many are unaware that boating while intoxicated is as serious an offense as driving an automobile while intoxicated. (Courtesy of author's collection, Tonto National Forest, permit CVC581.)

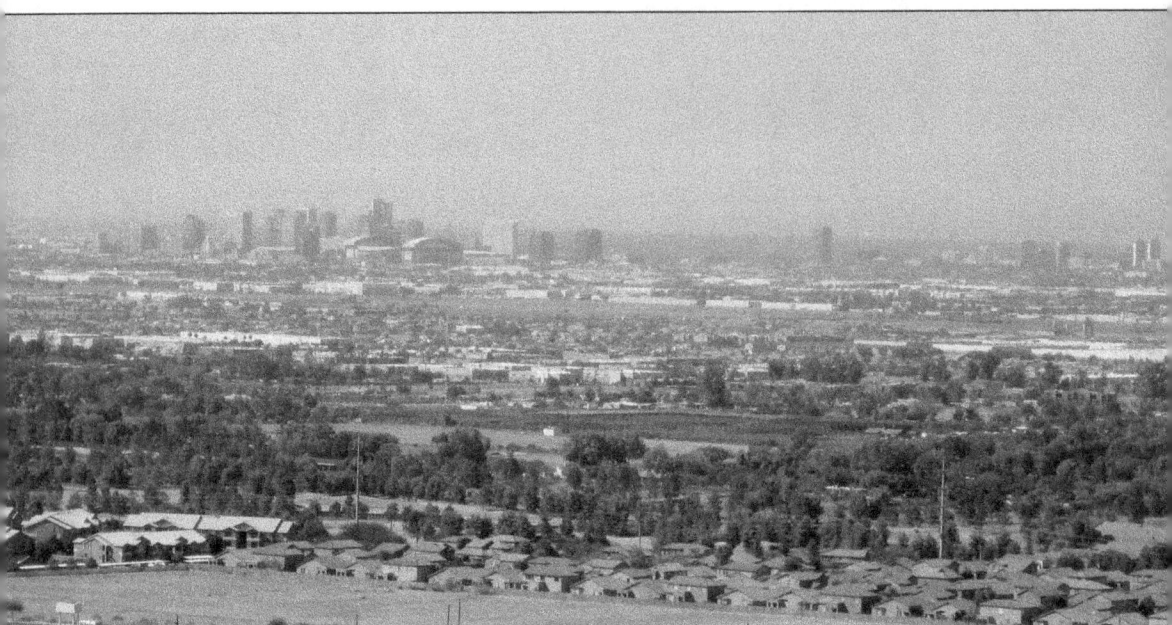

In 2009, Phoenix, Arizona, is part of a megalopolis that includes the cities of Mesa, Glendale, Scottsdale, Tempe, Peoria, Chandler, and Gilbert as well as Fountain Hills, Cave Creek, and the Yavapai Nation. Collectively known as the Valley of the Sun, it is completely dependent upon a meticulously altered environment anchored by several artificial lakes such as Bartlett and Horseshoe Dams on the Verde River. Most of the over four million people who live there simply take for granted that when they turn on their faucets, water will flow and life will go on. Many are unaware of genesis of the city's namesake, how like the mythical bird it is quite literally built upon the ruin of a people long gone. Whether it continues its growth unabated or, like the Hohokam, casts it ashes upon history's pyre will, as ever, be dependent upon its most valuable commodity—water. (Courtesy of author's collection.)

BIBLIOGRAPHY

Chaudhuri, Jonodev Osceola, ed. *The Yavapai of Fort McDowell : An Outline of Their History and Culture.* Mesa, AZ: Mead Publishing, 1995.

Giordano, Gerard. *Lake Pleasant.* Charleston, SC: Arcadia Publishing, 2009.

Hoffman, Teresa L., ed. *The Bartlett Reservoir Cultural Resources Survey. Cultural Resources Report No. 92.* Tempe, AZ: Archaeological Consulting Services, 1996.

Introcaso, David M. *Bartlett Dam, Verde River, Phoenix Vicinity, Maricopa County, Arizona: Photographs, Written Historical and Descriptive Data.* San Francisco: Historic American Engineering Record, National Park Service, Western Region, Department of the Interior, 1990.

Iverson, Peter. *Carlos Montezuma and the Changing World of American Indians.* Albuquerque: University of New Mexico Press, 2001.

Jackson, Donald C. and Clayton B. Fraser. *Historical American Engineering Record: Horseshoe Dam.* Loveland, CO: FRASERdesign, 1991.

Kupel, Douglas E. *Fuel for Growth: Water and Arizona's Urban Environment.* Tucson: University of Arizona Press, 2003.

Mason, Robert H. *Verde Valley Lore.* Scottsdale, AZ: L. J. Schuster Company, 1997.

Rogge, A. E., D. Lorne McWatters, Melissa Keane, and Richard P. Emanuel. *Raising Arizona's Dams: Daily Life, Danger, and Discrimination in the Dam Construction Camps of Central Arizona, 1890s–1940s.* Tucson: University of Arizona Press, 1995.

Speroff, Leon. *Carlos Montezuma, M.D.: A Yavapai American Hero: The Life and Times of an American Indian, 1866–1923.* Portland, OR: Arnica Publishing, 2003.

Verde River Water Project of the City of Phoenix, Arizona: A Vast Accomplishment of Lasting Importance. Phoenix: City of Phoenix, 1922.

Whittlesey, Stephanie M., Richard Ciolek-Torrello, and Jeffrey H. Altschul, eds. *Vanishing River: Landscapes and Lives of the Lower Verde Valley: the Lower Verde Archaeological Project : Overview, Synthesis, and Conclusions.* Tucson: SRI Press, 1998.

Visit us at
arcadiapublishing.com

www.ingramcontent.com/pod-product-compliance
Lightning Source LLC
Chambersburg PA
CBHW050606110426
42813CB00008B/2475